Collaboration and Governance in the Emergency Services

Paresh Wankhade • Swetketu Patnaik

Collaboration and Governance in the Emergency Services

Issues, Opportunities and Challenges

Paresh Wankhade
Edge Hill Business School
Edge Hill University
Ormskirk, UK

Swetketu Patnaik
Lord Ashcroft International Business
Anglia Ruskin University
Cambridge, UK

ISBN 978-3-030-21328-2   ISBN 978-3-030-21329-9   (eBook)
https://doi.org/10.1007/978-3-030-21329-9

Cover illustration: Pattern © Melisa Hasan

This Palgrave Pivot imprint is published by the registered company Springer Nature Switzerland AG
The registered company address is: Gewerbestrasse 11, 6330 Cham, Switzerland

PREFACE

The context of global emergency services landscape is changing. The recent tragic events in London, Paris, Manchester and around the world have highlighted the difficult and challenging role played by the emergency services. Their swift and professional response draws universal praise but also raises issues around reduced funding levels and job-cuts to match heightened security threat. Several barriers hamper collaboration efforts between the emergency services.

As the book argues, there are significant shifts in the demand patterns for blue light services. Ambulance demand in the UK is growing at an annual rate of about 5 per cent. The police services are witnessing a reduction in recorded crime but are dealing increasingly with cases relating to cybercrime, child and sexual exploitation and mental illness. Fire services have seen a massive reduction in incidence of fire. However, these organisations continue to be performance-managed and target-driven, and current models of service delivery do not reflect these changes, severely impacting the capacity to collaborate. Such unprecedented period of change is accompanied by major legislative and political changes, especially in the UK.

In addressing the issues around governance and interoperability between the emergency services, we offer critical insights and empirical evidence including our own research. Given the scope of this Pivot volume, we have focused on few key themes, which in our view are critical to the success of the collaboration and governance agenda. These include trust, leadership, workface resilience and organisational culture(s). We sincerely hope that the discussion and analysis provided on the chosen themes will not only be of help to emergency service staff, managers and service leaders but also appeal in equal measure to students, academics, public managers and policy makers.

Ormskirk, UK   Paresh Wankhade
Cambridge, UK   Swetketu Patnaik
July 2019

ACKNOWLEDGEMENTS

This project would not have been completed without the help of many people to whom we would like to express our sincere gratitude. We would like to thank our publisher Palgrave Macmillan, and in particular Jemima Warren and Oliver Foster, for their continuous support and understanding in completing this project.

We would also like to acknowledge the patience, support and understanding of our respective families (Kavita, Gaurav and Divij; Swati and Saisha) to write up this book.

This book is about the wonderful people working in the emergency services who deal with emotional and stressful situations on a daily basis, with a smile. To this end, this book is intended to assist service leaders and senior managers to deal with important issues concerning trust, workforce resilience and culture, as highlighted in this book, and to come up with innovative approaches and solutions to address them.

Paresh Wankhade
Swetketu Patnaik

CONTENTS

ABOUT THE AUTHORS

Paresh Wankhade is Professor of Leadership and Management at Edge Hill University Business School, UK. He is the programme leader for the UK's first bespoke Professional Doctorate in Emergency Services Management. He is the editor-in-chief of *International Journal of Emergency Services*. His research and publications focus on analyses of strategic leadership, organisational culture, organisational change and interoperability within the public services, with a focus on blue light services. Professor Wankhade has published in major journals, including *Work, Employment and Society, International Journal of Management Reviews, Public Management Review, Regional Studies, Public Money and Management* and *International Journal of Public Sector Management*. His recent work (co-authored with professionals) has explored leadership and management perspectives in the ambulance, police and fire and rescue services. He is currently working on a funded project to investigate sickness absence in the English NHS ambulance services.

Swetketu Patnaik is Senior Lecturer in International Business at Anglia Ruskin University, Cambridge, UK. He attained his PhD from the University of Liverpool, exploring dynamic evolution of inter-organisational collaborations in the development of new bio-pharmaceutical drugs. His research interests revolve around dynamic change of social systems and managerial endeavours to adapt. He particularly examines how practitioners experience and cope with challenges resulting from inter-organisational collaborations. He has recently concluded a 4-year-long longitudinal qualitative research on an ongoing strategic collaboration between two police forces in England. His work has been published in various journals including *International Journal of Knowledge Management* and *International Journal of Manpower*.

LIST OF FIGURES

LIST OF TABLES

Introduction and Background to Collaboration and Governance of Blue Light Emergency Services

Abstract This chapter sets the scene for this volume on the collaboration and governance between emergency services. It provides some justification behind this book and the rationale behind the choice of the key themes covered in this volume. The chapter first sets out the changing dynamics of the emergency services architecture and the shifts happening in the three main services—police, ambulance and fire and rescue. It then details the understanding of 'interoperability' and the difficulties in defining the concept. The Joint Decision Model and the JESIP (Joint Emergency Services Interoperability Programme) principles in the UK are then discussed, highlighting some of the latest challenges. The current legislative and governance framework is then detailed to keep the readers updated. The aims of this book are then discussed, highlighting the limitations of broader coverage within the scope of this book. This is followed by a plan of this volume, with a brief summary of each of the chapters. It is further argued that this volume is likely to appeal to a wider audience of emergency services staff, managers and leaders, policymakers, academics, scholars and researchers who are interested in management understanding of these important public services.

Keywords Emergency services • Police • Ambulance • Fire and rescue • Collaboration • Governance • Aims and scope • Limitations • Future research

© The Author(s) 2020
P. Wankhade, S. Patnaik, *Collaboration and Governance in the Emergency Services*,
https://doi.org/10.1007/978-3-030-21329-9_1

INTRODUCTION AND BACKGROUND

Management understanding of the emergency services and their role in dealing with the safety and preventative agenda in the society is on the rise (Wankhade et al. 2019). While the operation of the main blue light services, notably the ambulance, police and fire and rescue service, is quite global, there are huge variations and differences in the organisation, management and funding of these organisations including their service delivery models. Three challenges particularly merit some mention. For instance, in the UK, a large part of the blue light delivery resides in the public sector, funded by the government, with the three main services operating quite independently with different organisational and governance mechanism under respective ministerial insight. In Europe, the provision is much more fragmented, with a variety of public-private partnership models in vogue. In North America, the fire and rescue services (FRSs) are often the first responders to the emergency 911 calls, with the paramedic crews (and sometimes police) often part of the same team. This makes a systematic understanding of these organisations much more difficult. The second challenge stems from a dominance of professional/ practitioner literature, with pockets of academic knowledge developing in some aspects of the emergency work such as the police and paramedics, but a clear divide persists between the academic endeavours and professional knowledge, resulting in little co-production (Wankhade et al. 2019; Wankhade and Murphy 2012). The third challenge points to a climate of fiscal and budgetary pressures within which most of the provision of blue light service delivery happen, especially in Europe, North America and Australia. This book aims to address this gap and attempts to analyse the evidence from operation of, and governance framework for, the emergency services in the UK, often cited as the 'best practice' models around the world.

The global security climate and the recent tragic events in London and Manchester in the UK have highlighted the challenging role played by the emergency services in crisis situations. Their swift and professional response has drawn universal praise but has also raised issues around reduced funding levels and job-cuts to match heightened security threat. The horrific fire in the Grenfell Tower, a West London residential tower block in June 2017, resulting in a tragic loss of life is a defining moment for the fire and rescue services in the UK. The Grenfell Tower Inquiry led by Sir Martin Moore-Bick will be publishing its finding later this year about the response of the fire services which has drawn criticism. However it has also been reported that the "firefighters fear they are being 'stitched up' in the Grenfell Tower inquiry because their role has already been heavily scrutinised yet conclusions about the fire's causes are not likely to be drawn until at least three years after the disaster that claimed 72 lives" (Booth 2019).

Structural and cultural barriers hamper better collaboration and coordination of work between the emergency services (Parry et al. 2015). The call for 'transformational' reforms in the emergency services in the UK, particularly in this period of austerity that emphasises 'doing more with less', has been made elsewhere (Wankhade 2017). The 'transformational reforms', in essence, underpin the fundamental shift in the nature of the work and staff deployment across the three main blue light services—the ambulance, police and fire and rescue services. Significant shifts in the demand patterns for blue light services have been observed over the last decade. The ambulance services are witnessing an annual increase of 10% in the demand for 999 emergency calls (National Audit Office NAO 2017, 2011). Consequently, the service is struggling to meet its performance targets with available resources which is further affecting the health and well-being of the workforce (Wankhade and Mackway-Jones 2015; Granter et al. 2019; Wankhade et al. 2018).

The police services are also confronted with a different crime profile such as cybercrime, child and sexual exploitation, mental illness, in addition to tackling knife-, gun- and drug- related cases and community policing (Wankhade and Weir 2015). The College of Policing (2015) did some analysis of estimating demand on the police services and came up with interesting findings:

- Incidents involving people with mental health issues appear to be increasing;
- Demand on the police associated with protective statutory requirements, such as Multi-Agency Public Protection Panels, appears to be increasing; and
- Crime complexity has changed—cases of sexual abuse and cybercrime have been on increase too.

We see two indications of emerging pressure on police resilience. The first is the decreased levels of police visibility (community policing) and the second one is the increasing requests for mutual aid (NAO 2015a).

The fire and rescue services, over the past ten years, have witnessed a massive reduction in incidence of fire and are now required to look for creative options to utilise their workforce in building closer ties with the local ambulance and police services (Murphy and Greenhalgh 2018; National Audit Office (NAO) 2015a, b; Bain 2002). The House of Commons Public Accounts Committee HC PAC (HC 582) published its report (2016) on the financial sustainability of fire and rescue services. It commended the fire and rescue authorities to absorb funding reductions since 2010, but highlighted that risks to their financial and service sustainability could increase given the government's decision to implement further funding reductions from 2016 to 2017. Clearly, fire and rescue services need to make a strategic choice of either confining itself to putting off fires and lose paid staff or use its brand and engage in more boarder consultation and preventive work in collaboration with other public services (Mansfield 2015).

The collaboration agenda is further impacted by a lack of clear direction of travel, financial and budgetary challenges, differences in training and curriculum of staff across the services. The agenda is further impacted by the rise in issues such as mental health, stress and post-traumatic stress disorder (PTSD) in emergency services staff (MIND 2016; Donnelly 2017; Drury 2016; Gerber et al. 2010; Brough 2005).

The chapter is organised as follows. We first discuss the meaning and understanding of interoperability. We then analyse the current governance and oversight mechanism for the emergency services, using the UK as an example. The plan of this book is discussed next, followed by some concluding comments.

Defining Interoperability

The dictionary meaning of interoperability is "the ability of a system or component to function effectively with other systems or components" (Collins Dictionary 2018). Within the context of the emergency services, interoperability is generally understood as a multi-agency cooperation between them on issues around people, technology and resources. The need for such interoperability stems not only from the changing nature of the threats to national security but also on account of acute public service budget pressures for delivering safe and high-quality levels of services.

In official documents, interoperability is defined as "the capabilities of organisations or discrete parts of the same organisation to exchange operational information and to use it to inform their decision making"

(National Police Innovation Agency 2009). In the UK, the interoperability framework functions through the Joint Emergency Services Interoperability Programme (JESIP) which has been established to address the recommendations and findings from many major incident reports (www.jesip.org). It offers the joint interoperability training to the operational and tactical commanders from the three services (ambulance, fire and police) in preparing them to deal with initial response at a major or complex incident. The statutory meaning of what constitutes an 'emergency' including the agencies responsible for responding to such situations is detailed in the Civil Contingencies Act, 2004.

The UK government's latest strategy for counterterrorism (CONTEST 2011) further emphasises the drive to work together to improve interoperability by seeking to understand the different management and procedures (Donahue 2004), technology (Bevan and Hamblin 2009) and professional sub-cultures (Wankhade 2012), with the current economic climate providing a driver for joint working.

PROBLEMS OF DEFINITION AND SCOPE OF INTEROPERABILITY

There are several challenges to the general understanding of what interoperability actually means and what it covers? To begin with, the meaning is neither entirely clear nor consistent: in some cases it refers to the three blue light services (police, fire and ambulance). In other cases, it refers to all Category 1 responders (such as local councils) as defined by the Civil Contingencies Act (CCA) 2004. It may also however include Category 2 responders such as the volunteer agencies, utility companies or the community responders. To further add to the confusion, the Local Resilience Forums (LRFs) are required by the CCA 2004 to bring some of these organisations together. The official meaning follows the principles of INTEROPS, namely, Integration, Normal business, Training, Engagement, Resilient communications, Operationally acceptable, Planning and Sustainability (NPIA 2009). In our view, such an approach highlights the importance of operational procedures and compatible technology rather than the drivers and enablers of interoperability including the people and processes (Cole 2010a, b).

The complexity of the emergency community makes an objective assessment of the situation more difficult. The structure for managing the local multi-agency response to emergencies is also based on the Civil Contingencies Act (2004). For example, the Pitt Review (Pitt 2008), which looked at the response of the summer floods, highlighted that the overall response covered 28 fire, police and ambulance services; 28 community groups; 13 voluntary organisations; 86 local government organisations; 32 utility and infrastructure companies; 33 central government departments and agencies; 10 regional government offices and 43 Local Resilience Forums, amongst other agencies involved.

Lack of framework for national resilience also acts as a barrier to effective collaboration. Unlike the USA which has a federal response agency—the Federal Emergency Management Agency or FEMA, the fragmentary nature of the emergency management community in the UK further imposes additional barriers to closer interaction between different organisations. For instance, the three main blue light services work under different government departments and ministers. The Ambulance Services function under the Department of Health, the Police Service under the Home Office and the Fire and Rescue Service until recently (before January 2017) were under the jurisdiction of the Department of Communities and Local Government (DCLG) before being moved to the Home Office. Further, there are 52 police forces, 58 fire and rescue services and 12 NHS ambulance trusts in the UK. The sector fragments further if we consider the specialised agencies. For instance, the Environment Agency is part of the Department of Food and Rural Affairs (Defra), whereas the Coastguard Agency is an executive agency of the Department of Transport.

Review of literature suggests a rather confusing picture with competing models and frameworks. A 'hard' option on the lines of a single central agency at national level (e.g. Scottish Resilience, the FEMA in the USA) is often mooted with all its intended benefits such as having a national security agency, a national resilience forum and a national operations centre (legacy of Olympics facility) with a national resilience budget and procurement. This will allow greater consistency and standardisation across training, exercise and operating procedures. For example, the current system of gold/silver/bronze commanders operates differently for the three services and a more joined-up approach will prevent 'silo' thinking without any underpinning doctrine or strategy. On the other hand, a 'soft' approach on the lines of the current framework can also have its own strength to make a meaningful change. For instance, the local resilience forums (LRFs) can be strengthened with permanent staffs, ring-fenced budgets and powers to mandate training/equipment standards. Currently each individual organisation has its unique business model and unique expertise, and the arrangements is not a partnership of equals since initiatives focus on different organisations and multiple levels of agencies with differences in responsibilities, seniority and budgets. Our view is that interoperability depends upon three separate but connected elements of people, process/systems and technology and a 'one size fits all' approach is not always desirable, and the framework should allow flexibility to respond to local issues while keeping the bigger strategic picture in mind.

JESIP Principles

The JESIP principles serve as the basis of joint training and response to a major incident by the three main emergency services in the UK. Five key principles form the basis of such a response (see Fig. 1.1). These include:

- **Co-locate** with other services as soon as possible near the scene of the incident,
- **Communicate** in a clear language to build situational awareness,

Co-locate
Co-locate with commanders as soon as practicably possible at a single, safe and easily identified location near to the scene.

Communicate
Communicate clearly using plain English.

Co-ordinate
Co-ordinate by agreeing the lead service. Identify priorities, resources and capabilities for an effective response, including the timing of further meetings.

Jointly understand risk
Jointly understand risk by sharing information about the likelihood and potential impact of threats and hazards to agree potential control measures.

Shared situational awareness
Shared Situational Awareness established by using METHANE and the Joint Decision Model.

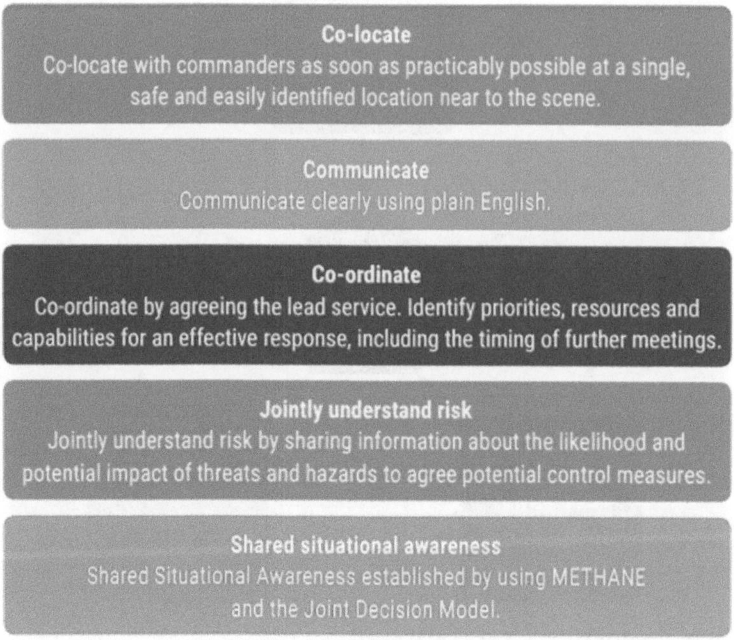

Fig. 1.1   JESIP principles. Source: www.jesip.org

- **Coordinate** the response to the incident including the lead organisation,
- **Jointly understand risk** by assessing the potential impact of the threat and
- **Shared situational awareness** to be established using other JESIP tools.

Joint Decision Model

Emergency services also make use of the JESIP Decision Model (JDM) while dealing with major incidents. This is a tool which assists different agencies involved in planning and reconciling different priorities and options (see Fig. 1.2). The core philosophy behind the JDM is about working together to save lives and reduce the threat and harm, likely to be

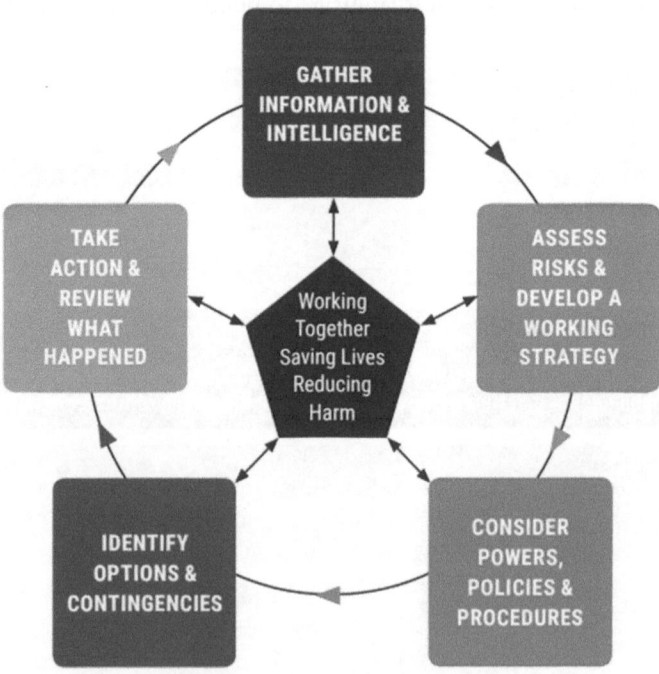

Fig. 1.2 JESIP Decision Model. Source: www.jesip.org

caused as a result of a major incident. It is a conceptual model involving five stages—information or intelligence gathering, assessing risk and developing strategy, considering policies and procedures, identifying options and taking actions. Being dynamic in nature, the model takes into account live information available for the incident to decide the best course of action. The JDM forms the basis of the decision-making for dealing with major incidents by emergency services in the UK.

In order to facilitate and drive collaboration efforts between the emergency services in England and Wales, the Emergency Services Collaboration Working Group (ESCWG) was set up in 2014. The working group comprises the senior leadership forums from the three services and the cross-party representatives.

EVIDENCE OF JESIP ON GROUND

A major report commissioned by the UK government to identify lessons learned from the response to major incidents since 1986 concluded that the "lessons identified from the events are not being learned to the extent that there is sufficient change in both policy and practice to prevent their repetition" (Pollock 2013). As part of the internal review, JESIP assurance visits were carried out during 2017 (JESIP Assurance Programme Report 2017). The visits were specifically "intended to measure the progress of services with embedding JESIP into their business as usual arrangements" (p. 4). While acknowledging that some good progress has been made by services in embedding JESIP, the report has also identified several concerns around its operation including (1) reliance on an individual or single department to deliver JESIP; (2) non-delivery of relevant JESIP elements across the organisation; (3) JESIP seen as only for major incidents and is a hindrance to JESIP becoming embedded across the emergency services and making it mainstream business as usual activity; and (4) JESIP is primarily a cultural change programme which needs embedding within the emergency services.

More recently, the Kerslake Independent Inquiry Report (2018) which examined the response of the emergency services on the Manchester Arena attack in May 2017 identified "lack of coordinated response and sharing of communication between the different organisations" including the failure of the National Mutual Aid Telephony system (p. 8) as major aspects of learning.

CURRENT COLLABORATIVE AND GOVERNANCE ARRANGEMENTS

As highlighted in the earlier sections, the collaborative framework between the emergency services in England and Wales is complex and each of the three main services is structured differently largely as a result of the relatively ad hoc nature of their historical development. The fire and rescue service (FRS) is accountable to the local Fire and Rescue Authority, which is made up of members appointed by the local authorities for the area covered by the FRS. The Authority produces an Integrated Risk Management Plan (IRMP) and monitors the budget of the service against that plan and other measures. The police services until recently had a 'tripartite relationship' as it has been called—between each Police Authority, the Chief Constable and the Home Secretary (Wankhade et al. 2016). The accountability of chief constables to the Home Secretary has been, inter alia, for the protection of citizens against national threats, while that to the Police Authority has been primarily in relation to local policing priorities. Through the advent of elected Police and Crime Commissioners (PCCs) in 2012, there has been a more direct accountability relationship that the electoral process implies between the Commissioners and their local public, but the operational experience has been not without controversy (Raine 2015; Raine and Keasey 2012; Lister 2013, 2014; Mawby and Smith 2013; Davies 2014; Wells and Smithson 2015). Scotland on the other hand has a national police force. The ambulance trusts in England are the first point of contact for providing emergency care to people with acute illness or injury by dialling either 999 or 111 emergency telephone numbers. When the National Health Service (NHS) was created in 1948, the ambulance service was a function given to the local authorities. The ambulance services are commissioned by the primary and acute trusts and work under the overall control of the Secretary of State who is responsible for the running of the Department of Health. Since becoming a part of the wider NHS in 1974, ambulance services have made huge progress, developing from a simple transport service into a mobile pre-hospital healthcare provider (Wankhade et al. 2015, 2018). There is a single ambulance model in Scotland and Wales.

The UK government's policy towards blue light collaboration, though a welcome development, is driven from the 'top' rather than pursing a 'bottom-up' approach. An *Emergency Services Working Group* was formed in 2014, supported by the Department of Health, Local Government and the Home Office to provide strategic leadership and coordination for improving the emergency services collaboration (Kane 2018). The government's 2015 manifesto made a commitment to enable fire and police service to 'work more closely together'. The Conservative Government in 2015 published the consultation paper, seeking views about the various options for closer integration between the emergency services (HM Government 2015). This was followed by publication of a response document in January 2016 outlining the government's position on measures to 'improve efficiency, building capacity and enhancing public confidence' (HM Government 2016). It introduced the Policing and Crime Bill in the Parliament in February 2016 which received the royal assent in January 2017. The Act now places a 'duty on police, fire and ambulance services to work together'. The Act also allows the elected Police and Crime Commissioners (PCCs) to now make a business case to run the police and fire and rescue services jointly, but ambulance services remain outside the purview of these provisions. One could argue that it is also worth exploring other means of promoting collaboration rather than choosing a legal duty. For example, financial incentives might motivate sharing of back office staff. It would be misleading for our readers to think that collaboration can only be achieved through legal or statutory duty. This has been happening at local level and there are numerous examples of good collaboration (ESCWG 2017; ESCWG 2016).

Another landmark change introduced in 2017 pertains to the creation of a new inspection body for the police and fire and rescue services—*Her Majesty's Inspectorate of Constabulary and Fire and Rescue Services* (HMICFRS). The erstwhile body responsible for independently assessing the police services (Her Majesty's Inspectorate of Constabulary, HMIC) was given the added responsibility of the fire services. An inspection programme for all fire and rescue services in England has been drawn up to assess the services on their 'effectiveness, efficiency and how well they look after their people'. Fire services are being judged as 'outstanding, good, requires improvement or inadequate' based on inspection findings, now made available for the first round of such annual inspections (see HMICFRS 2019). While the results are still being analysed, this is for the first time in their history that fire services have been independently assessed, which will have far-reaching significance for the future of these organisations and how they are managed and led.

We have argued here and in subsequent chapters that the current proposals are likely to help drive collaboration if attendant issues surrounding differences in organisational and professional cultures, differences in career spines and joint training programmes in the police and fire and rescue services at all levels are responded to. In the subsequent chapters we will examine some of these issues in greater detail, especially the issue of 'integrative' leadership, trust, workforce well-being and professional cultures in these important services, which are often neglected in management research.

PLAN OF THE BOOK

Chapter 2 explores the theoretical underpinning of the phenomenon of inter-organisational collaborations. Two theoretical perspectives, resource-based theories and institutional theory, were adopted to explore the rational and nature of partnerships. The chapter then provides an overview of the key themes or facets in relation to the broad body of inter-organisational relationship literature. The chapter then focuses on life cycle and processual evolution of inter-organisational collaborations. The chapter forwards the argument that strategic partnerships are not mere transaction relationships but are social constructs, and hence a better understanding of the social interaction could provide insights on how to manage these complex phenomena.

In Chap. 3, we focus our attention on two overlapping constructs—collaborative capabilities and collaborative leadership—and explore their implications on managing inter-organisational relationships. First, we explore the concept of collaborative capabilities and their constituent elements, including coordination, communication and interpersonal bonding and the implications for the blue light organisations. Then the chapter focuses on contemporary leadership theories and evaluates their implications on collaborative contexts. Current debates on shared and integrative leadership are particularly critiqued.

Chapter 4 explores the concept of inter-organisational trust. First, it defines and identifies the multidimensional nature of trust, particularly in collaborative relationships. In this context, specific attention is paid to trust-control and risk relationship. The chapter then highlights the dynamic nature of trust and argues that repeated pattern of interaction between the key individuals associated with the collaborations shapes the quality and development of trust. The chapter then explores the concept of psychological contract that signifies the trust relationship between employees and employers. The chapter makes the case for greater focus and consideration of psychological contract in the context of blue light collaborations because quality of psychological contract has an implication on workforce motivation and well-being, knowledge sharing and coordination of collaborative activities.

In Chap. 5, workforce issues are tackled. The chapter first analyses the current evidence on staff sickness and well-being in the three emergency services workforce, analysing the available evidence on the current sickness absence levels in the three services. The chapter then provides discussion on the growing number of cases of harassment and bullying of the emergency services workforce. The chapter details the blue light 'health and wellbeing frameworks' which are currently being developed and used by the emergency services along with other initiatives to help workforce well-being and resilience.

Chapter 6 deals with the important issue of professional 'culture(s)' in the three main emergency services. It provides a critical analysis of the changing identities of the workforce in a dynamic environment which is accompanied by changes in demand and service delivery in these organisations. The chapter begins with a review of the literature on 'organisational culture and change'. It then provides a discussion on the growth and evolution of the organisational culture(s) in the three services, drawing on evidence cited in the literature including policy documents. It sets out some conclusions on the issue of professional cultures while highlighting implications for the professionalisation agenda and interoperability issues between these services.

Chapter 7 is the concluding chapter drawing on the key findings from the themes discussed in this volume. It highlights opportunities and challenges emerging from our analysis of the themes. We also offer some suggestions for future research opportunities and renew our call for further empirical work on management of the emergency services by scholars, practitioners and management researchers.

CONCLUSION

This book is a timely review of the current blue light architecture and the integration drive of the UK government. It is our understanding that the 'blue light integration' project of the government is likely to succeed and facilitate greater degree of collaboration only if attendant issues surrounding differences in collaborative arrangements, organisational and professional cultures, leadership styles and trust and confidence issues are acknowledged and dealt with (Wankhade 2017). Another report (Parry et al. 2015, p. 27) on the evaluation of the emergency services has acknowledged that while "collaboration has been achieved in a number of ways and with a range of participants, it is 'patchy' and there is no 'one model' and it works on an area by area basis that reflects local need".

This book aims to highlight some of the key issues and address some of these key themes in greater details. It is topical, timely and provides a critical academic scrutiny of the opportunities and challenges the emergency services face in the 'new normal world'. It is aimed at the professionals, policymakers and academic scholars having interest on the evolution of the domain of the emergency services in the UK in general and on collaboration and governance of the emergency blue light services in particular.

LIMITATIONS AND FURTHER RESEARCH IMPLICATIONS

There are a few limitations about the scope and coverage in this short Pivot book. We are conscious of the decisions we have taken about the choice of the themes covered in this volume, but we feel confident that they are representative of the issues impacting the important agenda of collaboration and governance in the emergency services. Our choice also reflects opportunity for further empirical research on the chosen themes, and opportunities for further reflection by the professionals and emergency service personnel in the field. We have also tried to bring international perspectives wherever possible since a more detailed coverage was not possible. We however leave to our readers to judge whether we were correct in our thinking and approach.

This volume however provides several opportunities for a fruitful research agenda. Emergency services score high on the public surveys, but the issue of 'trust' remains controversial. Similarly, the notion of 'heroic' leadership is slowly diminishing and more distributed and 'integrative' models of leadership are also developing which can be explored with more empirical research. The mental health and well-being of emergency service workers remain a major concern and need further investigation across different settings. More cross-cultural case studies can also help to unpack the important but problematic issue of organisational culture(s) and its usefulness in understanding the changing professional identities of the emergency workers. We hope that this volume triggers greater interest in the understanding of these important public services, which are still underrepresented in the management research.

REFERENCES

Bain, G. (2002). *Independent review of the fire service*. London: Home Office.

Bevan, G., & Hamblin, R. (2009). Hitting and missing targets by ambulance services for emergency calls: Effects of different systems of performance measurement within the UK. *Journal of the Royal Statistical Society, 172*(Part 1), 161–190.

Booth, R. (2019, February 18). Firefighters worry they are being 'stitched up' by Grenfell inquiry. *The Guardian*. Retrieved February 18, 2019, from https://www.theguardian.com/uk-news/2019/feb/18/firefighters-worry-being-stitched-up-by-grenfell-inquiry.

Brough, P. (2005). A comparative investigation of the predictors of work-related psychological well-being within police, fire and ambulance workers. *New Zealand Journal of Psychology, 34*(2), 127–134.

Cole, J. (2010a). *Emergency service interoperability in jeopardy due to lack of standardised operational procedures*. Royal United Services Institute. Retrieved from http://www.rusi.org/go.php?structureID=commentary&ref=C4C330F4461949.

Cole, J. (2010b). *Interoperability in a crisis 2: Human factors and organisational processes*. Royal United Services Institute. Occasional Paper. Retrieved from http://www.rusi.org/downloads/assets/Interoperability_2_web.pdf.

College of Policing. (2015). *College of policing analysis: Estimating demand on the police service*. Coventry: College of Policing. Retrieved November 10, 2018, from https://www.college.police.uk/News/College-news/Documents/Demand%20Report%2023_1_15_noBleed.pdf.

Collins Dictionary. (2018). *Definition of interoperability*. Retrieved from https://www.collinsdictionary.com/dictionary/english/interoperability.

CONTEST. (2011). *The United Kingdom's strategy for countering terrorism*. London: Stationery Office.

Davies, M. (2014). The path to police and crime commissioners. *Safer Communities, 13*(1), 3–12.

Donahue, A. K. (2004). The influence of management on the cost of fire protection. *Journal of Policy Analysis and Management, 23*(1), 71–92.

Donnelly, L. (2017, February 4). Scandal-hit ambulance trust launches investigation into bullying and harassment claims. *The Telegraph*. Retrieved from http://www.telegraph.co.uk/news/2017/02/14/scandal-hit-ambulance-trust-launches-investigation-bullying-harassment/.

Drury, I. (2016, May 25). Theresa May slams fire service chiefs for allowing 'bullying and harassment' to flourish as she unveils sweeping reforms. *The Daily Mail*. Retrieved from http://www.dailymail.co.uk/news/article-3609247/Theresa-slams-fire-service-chiefs-allowing-bullying-harassment-flourish-unveils-sweeping-reforms.html#ixzz53V3qMvE7.

Emergency Services Collaboration Working Group. (2016). *National overview-2016*. Retrieved November 20, 2018, from https://aace.org.uk/wp-content/uploads/2016/11/National-overview-v13-WEB.pdf.

Emergency Services Collaboration Working Group. (2017). *Emergency services collaboration: The duty to collaborate-an information and support document*. Retrieved November 18, 2018, from https://www.nationalfirechiefs.org.uk/write/MediaUploads/NFCC%20Guidance%20publications/Operations/ESCWG_Guide_to_the_DtC.pdf.

Gerber, M., Hartmann, T., Brand, S., Holsboer-Trachsler, E., & Pühsea, U. (2010). The relationship between shift work, perceived stress, sleep and health in Swiss police officers. *Journal of Criminal Justice, 38*(6), 1167–1175.

Granter, E., Wankhade, P., McCann, L., Hassard, J. and Hyde, P. (2019). Multiple Dimensions of Work Intensity: Ambulance as Edgework. *Work. Employment and Society, 33*(2), 280–297.

Her Majesty's Government. (2015, September). *Consultation: Enabling closer working between the emergency services*. London: HM Government. Retrieved July 20, 2018, from https://www.ddfire.gov.uk/sites/default/files/attachments/Item%206%20Appendix%20A%20Consultation%20Enabling%20closer%20working%20between%20the%20Emergency%20Services.pdf.

Her Majesty's Government. (2016, January 26). *Enabling closer working between the emergency services: Summary of consultation responses and next steps*. London: HM Government. Retrieved March 20, 2018, from https://assets.publishing.service.gov.uk/government/uploads/system/uploads/attachment_data/file/495371/6.1722_HO_Enabling_Closer_Working_Between_the_Emergency_Services_Consult....pdf

Her Majesty's Inspectorate of Constabulary and Fire & Rescue Services HMICFRS. (2019). *Fire and rescue service assessments*. Retrieved January 10, 2019, from https://www.justiceinspectorates.gov.uk/hmicfrs/frs-assessment/frs-2018/.

HMICFRS. (2019). Fire and rescue service inspections 2018/19 – summary of findings from tranche 1. Retrieved February 19, 2019, from https://www.justiceinspectorates.gov.uk/hmicfrs/publications/fire-and-rescue-service-inspections-2018-19/.

JESIP. (2017). *JESIP assurance programme report on findings*, November 2017. Retrieved July 17, 2018, from https://jesip.org.uk/uploads/media/Documents%20Products/JESIP_Assurance_Programme_Report.1.pdf.

Kane, E. (2018). Collaboration in the emergency services. In P. Murphy & K. Greenhalgh (Eds.), *Fire and rescue services: Leadership and management perspectives* (pp. 77–91). Cham: Springer.

Kerslake Report. (2018). *An independent review into the preparedness for, and emergency response to, the Manchester Arena attack on 22nd May 2017*. Retrieved from https://www.jesip.org.uk/uploads/media/Documents%20Products/Kerslake_Report_Manchester_Are.pdf

Lister, S. (2013). The new politics of the police: Police and crime commissioners and the 'operational independence' of the police. *Policing, 7*(3), 239–247.

Lister, S. (2014). Scrutinising the role of the police and crime panel in the new era of police governance in England and Wales. *Safer Communities, 15*(/1), 22–31.

Mansfield, C. (2015). *Fire works: A collaborative way forward for the fire and rescue service.* London: New Local Government Network (NLGN).

Mawby, R. I., & Smith, K. (2013). Accounting for the police: The new PCC in England and Wales. *Police Journal: Theory, Practice and Principles, 86*(2), 143–157, 145.

MIND. (2016). One in four emergency services workers has thought about ending their lives. *MIND,* 20th April 2016. Retrieved July 17, 2018, from https://www.mind.org.uk/news-campaigns/news/one-in-four-emergency-services-workers-has-thought-about-ending-their-lives/#.W5-OhehKjIU.

Murphy, P. and Greenhalgh, G. (2018). *Fire and Rescue Services: Leadership and Management Perspectives.* Springer: Switzerland.

National Audit Office. (2015a). *Financial sustainability of police forces in England and Wales.* Retrieved February 18, 2018, from www.nao.org.uk/wp-content/uploads/2015/06/Financial-sustainability-of-policeforces.pdf.

National Audit Office. (2015b). *Impact of funding reductions on fire and rescue services.* Retrieved February 18, 2018, from www.nao.org.uk/report/impact-of-funding-reductions-on-fire-and-rescue-services/.

National Audit Office NAO. (2011). *Transforming NHS ambulance services.* London: Stationery Office.

National Audit Office NAO. (2017). *NHS ambulance services.* HC 972, Session 2016-17. London: Stationery Office.

National Police Innovation Agency, NPIA. (2009). *Guidance on multi-agency interoperability.* London, UK: National Policing Improvement Agency.

Parry, J., Kane, E., Martin, D., & Bandyopadhyay, S. (2015). *Research into emergency services collaboration* (Emergency Services Working Group Research Project). Sheffield University.

Pitt, M. (2008). *Lessons learned from the 2007 floods.* London: Cabinet Office.

Pollock, K. (2013). *Review of persistent lessons identified relating to interoperability from emergencies and major incidents since 1986.* A report commissioned by the Cabinet Office Civil Contingencies Secretariat. Retrieved July 20, 2018, from https://jesip.org.uk/uploads/media/Documents%20Products/Pollock_Review_Oct_2013.pdf.

Raine, J. W. (2015). Enhancing police accountability in England and Wales: What differences are police and crime commissioners making? In P. Wankhade & D. Weir (Eds.), *Police services: Leadership and management perspectives* (pp. 97–114). New York: Springer.

Raine, J. W., & Keasey, P. (2012). From police authorities to police and crime commissioners: Might policing become more publicly accountable? *International Journal of Emergency Services, 1*(2), 122–134.

Wankhade, P. (2012). Different cultures of management and their relationships with organizational performance: Evidence from the UK ambulance service. *Public Money & Management, 32*(5), 381–388.

Wankhade, P. (2017, July 13). How to reboot Britain's fractured emergency services. *The Conversation*. Retrieved from https://theconversation.com/how-to-reboot-britains-fractured-emergency-services-79528.

Wankhade, P., & Mackway-Jones, K. (Eds.). (2015). *Ambulance services: Leadership and management perspectives*. New York: Springer.

Wankhade, P., & Murphy, M. (2012). Bridging the theory and practice gap in emergency services research: A case for a new journal. *International Journal of Emergency Services, 1*(1), 4–9.

Wankhade, P., & Weir, D. (Eds.). (2015). *Police services: Leadership and management perspectives*. New York: Springer.

Wankhade, P., Radcliffe, J., & Heath, G. (2015). Organisational and professional cultures: An ambulance perspective. In P. Wankhade & K. Mackway-Jones (Eds.), *Ambulance services: Leadership and management perspectives*. London: Springer.

Wankhade, P., Radcliffe, J., & Heath, G. (2016). *Coordination of emergency services and the problem of governance: A UK perspective*. European Academy of Management (EURAM) 2016 Conference, University Paris-Est Créteil (UPEC), Paris, France, 1 June–4 June 2016.

Wankhade, P., Heath, G., & Radcliffe, J. (2018). Cultural change and perpetuation in organisations: Evidence from an English Emergency Ambulance Service. *Public Management Review, 20*(6), 923–948.

Wankhade, P., McCann, L., & Murphy, P. (Eds.). (2019). *Critical perspectives on the management and organization of emergency services*. New York: Routledge.

Wells, H., & Smithson, H. (2015). Grey areas and fine lines: Negotiating operational independence in the era of the police and crime commissioner. *Safer Communities, 14*(4), 193–204.

CHAPTER 2

Theoretical Underpinnings of Collaborations Amongst Emergency Service Organisations

Abstract This chapter explores the rationale underpinning inter-organisational collaborations amongst blue light organisations. Resource-based theories suggest that collaborations are formed between organisations to overcome resource constraints, whereas institutional theory suggests that collaborations are a strategy that organisations adopt to conform to changing institutional environment. Although both theoretical perspectives provide rationale for the formation of these arrangements, they do not account for how, in practice, collaborations could evolve over time and what role do individuals play in shaping the transformation of the collaboration. In this respect, we suggest that adopting process perspective could shed more light on different dimensions underpinning development of inter-organisational relationships over time.

Keywords Blue light services • Collaborations • Resource-based theories • Institutional theory • Process theory

INTRODUCTION

Enhancing cooperation between the three blue light services to improve efficiency, interoperability and resilience is one of the key objectives of the UK government. From a structural point of view, all the three blue light

© The Author(s) 2020
P. Wankhade, S. Patnaik, *Collaboration and Governance in the Emergency Services*,
https://doi.org/10.1007/978-3-030-21329-9_2

services started to work together under the Joint Emergency Services Interoperability Programme (JESIP) that was established in 2012. JESIP came into existence following a report by the Association of Chief Police Officers, Chief Fire Officers Association and Association of Ambulance Chief Executives to improve collaboration between emergency services at the scene of major incidents. A major incident is defined as, "An event or situation, with a range of serious consequences, which requires special arrangements to be implemented by one or more emergency responder agencies" (JESIP Joint Doctrine 2016). Simply put, the nature of incident(s)/emergencies necessitates cooperation between the three emergency services.

Beyond major incidents and emergencies, collaborations between and amongst the emergency services are increasingly encouraged to be adopted as a strategy to achieve efficiency and reduce cost. In this context, Sir Ken Knight's *Facing the Future* (2013) review of the fire and rescue services was instrumental in making the case for collaborations amongst the fire services as well as with other local services. "Collaboration in all its forms", he asserts, "is the answer to improving the service, making services more interoperable and, of course, reducing duplication of spend" (*Facing the Future* 2013, p. 45). More recently, the Policing and Crime Act 2017 goes further and identifies collaboration as a priority for blue light services by allowing Police and Crime Commissioners to take control of Fire and Rescue Authorities. In other words, collaboration— between the different services as well as amongst different authorities— is considered as an integral part of producing and delivering effective blue light services.

Notwithstanding the significance attached to collaborations by the policymakers, empirical studies show that a vast majority of inter-organisational relationships are inherently unstable and suffer from high failure rates (Niesten and Jolink 2015; Kale and Singh 2009). Different researchers have reported that failure of collaborations range from as high as 70% (Harrigan 1988) to 50% (Duysters et al. 2012). Although most of these figures pertain to international collaborations between organisations from different countries, the challenges collaborations in public sectors face seem to be consistent to collaborations in the private sector. Collaborations are indeed associated with high cost of delivery, conflicts and tensions and inertia to the achieve any significant advantage in the long term (see for instance Bryson et al. 2006; Grimshaw et al. 2002; Huxham and Vangen 2005).

2 THEORETICAL UNDERPINNINGS OF COLLABORATIONS... 21

Therefore, it is imperative that we take a critical view of collaborations in general and in the context of emergency services in particular. In this chapter, we explore and delineate, broadly, the rationale, theoretical perspectives and critical issues relating to inter-organisational collaborations. This chapter is structured as follows. First, we define the phenomenon of inter-organisational collaborations and, in this respect, we focus our attention on the resource-based and institutional theories. Then we identify various dimensions and aspects relating to the phenomenon. Next, we pay attention to the life cycle and processual perspectives that underpin the dynamic evolution of inter-organisational relationships. We conclude the chapter with our reflection on implications of processual view on studying collaborations.

MEANING OF INTER-ORGANISATIONAL COLLABORATIONS

Although inter-organisational collaborations have emerged as the single most commonly adopted strategy in case of for-profit organisations (Gulati and Gargiulo 1999; de Rond 2003), these have also attained popularity amongst public service organisations (O'Leary and Bingham 2007; Bryson et al. 2006; Osborne 2009). The incidence of inter-organisational collaborations amongst public organisations is primarily driven by complex problems faced by public organisation, often underpinned by increasing demands with fewer available resources coupled with greater scrutiny on performance. It is often argued that adopting a joined-up approach to public policy and service-related issues would immensely contribute in delivery of services. O'Leary and Bingham (2007, p. 7) define collaborative public management as "a concept that describes the process of facilitating and operating in multi-organizational arrangements to solve problems that cannot be solved or easily solved by single organizations. Collaborative means to co-labour, to co-operate to achieve common goals, working across boundaries in multisector relationships. Cooperation is based on the value of reciprocity." In essence, collaborative approach to problem solving, in public services, is increasingly considered to increase "efficiency, flexibility and innovation, local adaptation and enhanced community ties" (Chen 2010, p. 382). Notwithstanding the need for collaborative approach to public management, it is equally critical to define inter-organisational collaboration and its critical salient features.

Inter-organisational collaborations are defined as, "close, long term, mutually beneficial agreements between two or more partners in which resources, knowledge and capabilities are shared with the objective of enhancing competitive position of each partner" (Spekman et al. 1998, p. 748). This encompassing definition includes various inter-organisational relationships such as joint ventures, research and development partnerships, multi-partner alliances and consortiums, public-private partnerships but excludes simple market transactions and mergers and acquisitions. This definition captures four salient features of strategic collaborations. First, collaborations, irrespective of their nature, pertain to compatible goals of each of the partner organisations. In essence, the partners aim to use their collaboration as a strategic instrument to achieve their respective strategic and operational objectives. Second, collaborations involve an element of engagement between two or more independent entities that seek to access each other's resources and capabilities. In the process, the partners become dependent on each other to realise their joint and individual objectives. Third, collaborations have an element of long-term commitment by partners. The commitment takes place in the backdrop of an absence of hierarchical governance (Provan and Kenis 2008), particularly so in the case of collaborations between public organisations. Fourth, collaborations are transitional entities that respective partners could easily dissolve at their convenient time. Simply put, collaborations have their respective life cycles and they evolve over time. Therefore, mitigating premature termination as well as achieving desired performance entails systematic management of structural and social contingencies, which as empirical research suggests has been proved to be extremely challenging.

Therefore, the following question needs deeper exploration—why do organisations form collaborations, when the rate of failure and underachievement is significantly high? This question, particularly in the context of public management, including emergency services, needs deeper theoretical exploration. Different scholars have explored the inter-organisational collaboration phenomenon through multiple theoretical lens.[1] Resource-based theories and institutional theory are, perhaps, the

---

[1] Faulkner and de Rond (2000), in a detailed review of inter-organisational relationship literature, identified six perspectives that have originated from economics and four perspectives which trace their origin to sociology in general and organisational theory in particular. The perspectives that originate from economics include market power theory, transactional cost theory, the resource-based view, agency theory, game theory and the real option theory, among others. Theoretical perspectives that trace their roots to sociology and organisational

most useful perspectives that provide theoretical understanding for formation of inter-organisational collaborations in emergency services.

RESOURCE-BASED THEORIES AND FORMATION OF INTER-ORGANISATIONAL COLLABORATIONS

Resource-based theories conceptualise organisations as possessing 'bundles of resources' and the capacity to configure these resources underpins the survival and success of the organisations. In this respect, Barney (1991) identifies resources as "all assets, capabilities, organisational processes, firm attributes, information, knowledge, etc. that are controlled by a firm that enable the firm to conceive of and implement strategies that improve its efficiency and effectiveness" (Barney 1991, p. 101). He organised resources under three broad categories, namely (a), physical capital resources that include tangible assets (e.g. land, plant, equipment, finished and semi-finished goods) and intangible assets (e.g. brand name, patents, copy rights, reputation); (b) human capital resources that pertain to educational background, skills experiences, relationships, intelligence of individual staffs in an organisation; and (c) organisational capital resources, such as organisational structure, rules, procedures and organisational culture as well as the firm's relationship with external institutions. Therefore, when organisations, public or private, are encouraged to adopt collaborative approach, there is acceptance that, notwithstanding the resources they possess, the organisations, on their own, cannot successfully solve complex problems. Thus, inter-organisational collaboration represents a strategic initiative to "access other firms' resources for the purpose of garnering otherwise unavailable competitive advantages and value ... in sum it is about creating the most value out of one's existing resources by combining these with others' resources, provided, of course, that this combination results in optimal returns" (Das and Teng 2000, p. 36/37).

In other words, complex set of problems, in essence, highlight an organisation's resource limitation, which it could address only by forming partnerships with other organisations. Access or acquisition of resources inevitably raises the question of 'power' in such relationships. Perhaps of all the organisational (and strategic) paradigms, resource-dependence theory has contributed the most in producing a comprehensive amount of 'power' within organisations, between organisations and in others in their

theory include resource-dependence theory, relational contract theory, organisational learning theory and social network theory.

external environment. In this context, Pfeffer and Salancik's (1978) semi-nal work *External Control of Organisations* analyses the source and conse-quences of power in inter-organisational relations. Resource-dependence theory posits the view that organisations cannot have access to all resources all the time and therefore are dependent on resources that reside outside their organisational boundaries. Hence, the motivation of those running the organisations is not merely to access resources to ensure the organisa-tions' survival but also to enhance their own autonomy, while at the same time maintaining stability in the respective organisations' exchange rela-tions. The most significant finding of Pfeffer and Salancik's work was that when it comes to explaining strategy, 'power' often dominates the need for efficiency and effectiveness, which are accepted as broad measures for organisational performance (Davis and Cobb 2010). Simply put, organisa-tions, to grow and survive, necessarily have to trade or exchange vital resources with other organisations in their environment. The resource dependence among organisations externally constrains their behaviour and choices and eventually creates power differentials between the firm and the elements of its environment. Therefore, in seeking recourses from other organisations, managers aim to fulfil two objectives: to reduce the power of the partner organisations and to bolster their own power over others within their own organisation in particular.

To summarise, resource-based perspectives sensitise us to (a) organisa-tional resources and the limitations of any organisation to possess all resources at all times, (b) the necessity to access and acquire resources from other organisations to co-develop and/or co-deliver products and services and (c) power relations that exist within organisations and how they manifest in inter-organisational context.

In specific context to emergency services, although the 'call to share resources' echoes loudest, evidence shows that, on their own, only a lim-ited number organisations really pursued collaborations with other organ-isations. Sir Knight in his extensive report on the current state of fire and rescue services '*Facing the Future*' (2013) found that although the Fire and Rescue Service Act 2004 enables the fire and service authorities to voluntarily explore ways to collaborate, there has been only one 'success-ful' collaborations between the fire and rescue authorities. He thus concludes that there is 'little appetite or incentives' to enter into collab-orative endeavours (Sir Knight 2013, p. 45). In contrast, by 2011, the police forces had identified and planned almost 543 collaborations with more than two-third of the projects involving collaborations between two forces (HMIC 2012, p. 5), though little is known on how they have fared.

Institutional Theory and Formation of Inter-Organisational Collaborations

Scholars who explore organisations actions through institutional perspective emphasise that organisational actions, such as forming alliances, are influenced by the 'institutional climate', which puts pressures on organisations to acquire legitimacy by adhering to socially justified actions or norms (DiMaggio and Powell 1983; Scott 1995). In other words, organisational actions, in essence, reflect embeddedness of organisations in socio-structural and normative contexts. Due to the embeddedness of organisations in their institutional context, firms seek legitimacy and approval from other organisations and constituents, through their actions (Dacin et al. 2007; Oliver 1996). Organisations also gain or increase legitimacy by adhering to institutional rules, regulations, norms and expectations which are directly or indirectly imposed by wider institutional environment (e.g. government, society, other players in the industry/sector, and community) (Scott 1995; Dacin et al. 1999).

Collaborations amongst the emergency service are high on the agenda of the Government of UK. For instance, HM Government in 2015 undertook consultation on how to facilitate greater collaboration and in fact how to make collaboration a 'legal duty' for the three emergency services. In this respect, the coalition government's consultation paper drew attention to Conservative's manifesto pledge that called for greater integration amongst the police and fire services. It read: "Our manifesto was clear that—we will enable fire and police services to work more closely together and develop the role of our elected and accountable police and Crime Commissioners" (HM Government Consultation Paper 2015). Simply put, the UK government, particularly in the absence of initiatives from local emergency services to voluntarily engage in collaboration amongst and with other services, becomes the key driver of the strategy as well as the process. The rationale for collaboration is not merely about sharing resources, but becoming more efficient and effective. In making the case for collaboration between and amongst emergency services, the consultation document states, "We know that collaboration presents a real opportunity for organisations in terms of increasing efficiency and effectiveness alongside the ever-present need to maximise available resources" (HM Government Consultation Document 2015, p. 6). Notwithstanding the emphasis on 'increasing efficiency and effectiveness', reduction in the overall expenditure, it seems, is the most significant underpinning for pursuing collaboration.

Collaborations between organisations, seen through institutional theory lens, are considered as a strategy to conform to changes in the institutional environment (Provan et al. 2007; Bryson et al. 2006). For instance in specific context to the police services, the October 2010 Comprehensive Spending Review (CSR) outlined a 20% cut in the central government police funding grant for all 43 forces in England and Wales by 2014/2015 (in real terms). Collaborations, in this context, were suggested as a viable strategy to mitigate the impact of this reduction in resources in real terms. HMIC 2012 report made a strong case for collaboration amongst police forces by highlighting that collaboration in police service is not a new phenomenon but resource sharing "can also result in significant saving. This makes collaboration—within with another force, public or private sector—one option available to the police as they work to close the 20% saving requirement outlined in the October 2010 Spending Review" (HMIC Report 2011, p. 4). Similarly, Sir Knight's review, *Facing the Future*, on the fire and rescue services also considered collaboration as a mechanism to achieve higher saving. At the same time, he also made a case for collaboration amongst the services more holistically and suggested that "collaboration in all its forms is the answer to improving the service, making services more interoperable and of course reducing duplication of speed" (p. 45). In this context, collaborations amongst emergency service organisations in the UK are considered as critical due to the changes in the overarching institutional logic, particularly during a period that characterises reduction of public sector spending and implementation of austerity programme.

OVERVIEW OF DIFFERENT ASPECTS RELATING TO INTER-ORGANISATIONAL COLLABORATIONS

Different scholars have organised research on inter-organisational relationships under various categories, which, in essence, highlights its various dimensions and aspects. For instance, Gulati (1998) segregated inter-organisational relationship literature under two categories: (a) key decisions that are common to all strategic collaborations irrespective of their types and forms and (b) performance consequences of partnerships. The first category of literature explores questions relating to (1) formation of collaborations (e.g. why [and which] organisations form strategic partnerships and with whom?); (2) governance structure of the collaborations (e.g. what types of formal contractual forms the partners adopt to formalise their relationships?); and (3) the dynamic evolution of inter-organisational relationships (e.g. how does the partnership and the roles of the partners evolve over?). The second category of literature focuses on questions relating to (4) performance of collaborations (e.g. what factors influence the success (or failure) of partnerships?) and (5) performance consequences for collaborating organisations (e.g. does the collaboration have an influence on the overarching performance of the organisations?). Building on the work of Gulati (1998) and Spekman et al. (1998), Patnaik (2011) organised the inter-organisational relationship literature around seven thematic categories, namely (a) the rationale for the formation of collaborations (McGuire 2006), (b) governance structure and mechanisms in collaborations (Ansell and Gash 2008; Bryson et al. 2006), (c) learning and knowledge in strategic partnerships (Powell 1998; Hibbert and Huxham 2010), (d) trust in collaborative context (Gulati 1998), (e) tensions and instability in collaborations (Das and Teng 2002; de Rond and Bouchikhi 2004), (f) performance of partnerships (Provan and Kenis 2008; Provan and Milward 2001) and (g) the dynamic evolution of inter-organisational relationships (Thomson and Perry 2006).

In critiquing the existing bodies of literature on these different facets, albeit in a different industry/sector context, Patnaik (2011) observed two common threads that run through all the aforementioned themes. First, all the themes are intertwined, in the sense that a specific theme cannot provide us with a better understanding unless it is explored in relation to the other themes. For example, how collaborations perform could be better explored in context to existing governance structure, levels of trust, existence of tensions and experiences of learning and knowledge sharing. Persistent call for more focus on studying the unfolding processes of the evolution of inter-organisational arrangements is the second common thread that runs through different thematic categories. This follows from the argument that the nature of alliance phenomenon not only is inherently complex but also follows a highly uncertain and evolutionary path. Hence, the calls for more studies to focus on the developmental processes to capture are not only about exploring the dynamic nature of collaborative governance structures, inter-organisational trust, knowledge transfer in partnerships and instabilities, but also to explain how these aspects manifest in an inter-organisational relationship and how they relate to one another as the collaboration evolves over time (Parkhe 1993; Doz 1996; Yan and Zeng 1999; Inkpen and Tsang 2007; Zaheer and Harris 2006; Ness 2009). Therefore, scholars have emphasised that the evolutionary path that inter-organisational relationships follow could have significant impact on their performance (Harrigan 1988; Lunnan and Haugland 2008) and therefore understanding the dynamic process of the evolution of inter-organisational collaborations could provide more critical insights on how such organisational arrangements could be better managed (Larson 1992; Ring and Van de Ven 1994; Spekman et al. 1998).

Thomson and Perry (2006), in their review of collaborations in the context of public administration in the United States, identify two competing political traditions, namely classic liberalism and civic republicanism, underpin our understanding of the phenomenon. They assert:

> classic liberalism, with its emphasis on private interest, views collaboration as a process that aggregates private preferences into collective choices through self-interested bargaining. Organisations enter into collaborative agreements to achieve their own goals, negotiating among competing interests and brokering coalitions among competing value systems, expectations and self-interested motivations. Civic republicanism, on the other hand with its emphasis on a commitment to something larger than the individual (whether that be a neighbourhood or the state), views collaboration as an integrative process that treats differences as the basis for deliberation in order to arrive at mutual understanding, a collective will, trust, and sympathy and the implementation of shared preferences. (Thomson and Perry 2006, p. 20)

The elucidation of the two competing roots is critical because, in essence, they provide philosophical underpinnings to practical complexities that characterise numerous challenges associated with the performances and survival of strategic partnerships. Thomson and Perry (2006) build on the work of Wood and Gray (1991), who asserted that "the doing of collaboration—the process component—is a black box" (Thomson and Perry 2006, p. 21, citing Wood and Gray 1991), and suggested that the complexities associated with collaborations could be better captured using the 'antecedent—process—outcome model'. The antecedent provides the context and rationale for formation of collaborations, and 'process' is a result of interaction between its five components, namely governance, administration, organisational autonomy, mutuality and norms of trust and reciprocity. Put simply, process is conceptualised as a moderator between the antecedent (conditions) and the outcomes (Chen 2010). This conceptualisation of process is problematic because it treats the antecedents and outcomes as distinct from the process. That apart, the model ignores two distinctive features of process. First, it broadly ignores the notion of time (Salk 2005), particularly after a partnership is formed, and second it completely overlooks social actors. Nevertheless, the model exhorts partnership managers to acknowledge and be sensitive of the five components of the process.

Although calls to study developmental processes underpinning inter-organisational relationship are not new, this stream of research is indeed under-represented.

Salk (2005) laments on the state of process studies on strategic alliances and notes:

> every so often, alliance scholars make cogent arguments for why the alliance field needs more process research. Rather than plea this case yet again, this paper explores why the field of alliance research continues to be overwhelmingly cross-sectional and structural in nature. Looking at research published since the mid-1990s, it appears that reflexive 'calls for future process research' crop up frequently. Even in cross-sectional, structural research, authors evoke stories about processes, even if these processes depend on assumptions about actors and behaviors that lie outside of what is observable with their data and methods. What impedes the development of process research are norms and taken-for-granted routines within the mainstream scholarly community. (Salk, 2005: 117)

Therefore, it is imperative that we develop a better understanding of how process is conceptualised and the contribution of the research on the evolution of the inter-organisational collaborations to effectively manage these complex social phenomenon

THE DYNAMIC EVOLUTION OF INTER-ORGANISATIONAL COLLABORATIONS

Studies on the dynamic processes that underlie the evolution of inter-organisational collaborations tend to focus on the process of formation, development and dissolution to explain how and why these arrangements undergo transformation over time (Ring and Van de Ven 1994; Doz 1996; de Rond and Bouchikhi 2004; Salk 2005). Contractor (2005) notes that by the term 'process', management scholars essentially refer to human interaction and behavioural dimension in organisational and managerial activities.[2] Therefore, studies on the evolution of inter-organisational partnerships seek to focus on the social relationships that underpin continued interaction between those who organise and manage alliances over time. The social nature of strategic alliances highlights the significance of social

[2] In this context Contractor also notes that the term 'structure' within alliance research refers to the "legal form of the relationship (e.g. licensing, supply chain contract, cross-shareholding etc.)" (p. 124).

relationships and interaction between key individuals involved in an inter-organisational collaboration, reciprocity norms, trust and an informal climate underneath the formal contractual and structural governance mechanism (Larson 1992; Powell 1998).

Although Reuer et al. (2002) trace the roots of research on the evolution of inter-organisational collaborations to the work of Franko (1971), who examined a multinational firm's evolving tolerance for international joint ventures to explain alliance instability, Van de Ven (1976) was amongst the first to call for systematically studying dynamic evolution of inter-organisational relationship. In his seminal paper titled "On the nature, formation and maintenance of relations among organisations", Van de Ven (1976), in the backdrop of increasing collaborations between public organisations in the USA, called for more research to study how these inter-organisational arrangements are formed and sustained. Building on resource-dependence perspective, he emphasised that organisations collaborate when they are not able to achieve their respective objectives on their own, and in the process of collaborating with one another, the organisations create a social system whose overarching objective is to attain goals that either of the partners on their own could not have been able to achieve. The newly created social system, according to Van de Ven (1976), not only adopts a structure which provides a broad administrative arrangement to establish role relationship amongst the members but also attempts to develop a process that could facilitate sharing of resources and flow of information between the members. According to him, "emergence and functioning of an inter-organisational relation is a cyclical process of: need for resources—issue commitments—inter-agency communications to spread awareness and consensus—resource transactions—and structural adaptation and pattern maintenance over time ... what starts as an interim solution to a problem may eventually become a long term inter organisational commitment of resource transaction if previous cycles in the process are perceived by the parties to have been effective encounters" (Van de Ven 1976, p. 33). Interestingly, the cyclical conceptualisation of process became popular only in mid-1990s and, until then, most scholars followed the seminal organisational life cycle model developed by Larry Greiner (1972).

LIFE CYCLE PERSPECTIVE

Inter-organisational collaboration as a strategy, particularly amongst public service organisations, started to take shape in 1980s (for reviews, see Galaskiewicz 1985; Oliver 1990) and D'Aunno and Zuckerman (1987) were amongst the earliest scholars who provided any empirical support on how the collaborations between public service organisations develop over time. They studied collaborative activities amongst individual hospitals to create hospital federations, and based on their in-depth research on the life span of one of the federations, they developed a four-phased life cycle model by focusing on the factors that they identify as those which influenced the evolution from one stage to another. The four stages through which alliances evolve include (a) emergence of coalition between two or more hospitals, predominantly due to environmental threats and resource uncertainty; (b) transition of the coalition into a federation, which entails hiring of a management team to oversee coordination of work and efforts; (c) maturity of the federation, which signifies continuation of investment of resources from partner hospitals in the federation as well as willingness of individual organisations to put the interests of the federation ahead of their own concerns; and (d) critical crossroads, which result due to increase in centralisation of decision-making in the federation, which in turn forces partner organisations to withdraw from the federation. Grey (1989) suggested a three-phase framework, namely (a) problem setting, (b) direction setting and (c) implementation. Forrest and Martin (1992) segregated the collaboration process into four stages: (a) matching stage, key elements of which include appropriate timing, finding the right partner and assessing the cost of forming a collaboration; (b) negotiating the collaboration, which may vary from six months to a year and during the process partners must spell out their expectations as well as detailed technical information; (c) developing the agreement by involving technical, business and legal personnel; and (d) implementation of the partnership, which would necessitate good leadership, sound interpersonal and communication skills and excellent commercial judgement. In their study they concluded that success in inter-organisational collaborations was dependent on the choice of the right partner and was facilitated by putting time and effort into negotiating the terms of the collaboration and appropriately implementing them.

Similarly, Murray and Mahon (1993) asserted that inter-organisational collaborations "experience a 'life cycle' of formation, development, maintenance and dissolution" (p.102) and comprise five stages, namely: (a) courtship, wherein partners investigate each other and evaluate each other's strength and weaknesses; (b) negotiation, based on each other's evaluation of potential benefits and costs as a result of entering into the alliance the partners arrive at contractual agreements; (c) start-up, which follows the negotiation stage and involve operation of joint activities; (d) maintenance, which characterise routinisation of joint operations; and (e) ending or dissolution. In early 1990s, Kanter (1994) undertook a detailed study involving more than 37 companies and their partners from more than 11 countries to understand why certain partnerships are more "productive" as compared to other partnerships (p. 96). Based on her findings she concluded that collaborations are more than mere transactional deals between two organisations, and argued that: "they *(collaborations)* are living systems that evolve progressively in their possibilities ... and which cannot be 'controlled' by formal systems but require dense web of interpersonal connections and internal infrastructures that enhance learning ... *and the successful alliances are the one that* ... involve collaboration (creating new value together) rather than mere exchange (getting something back for what you put in)" (Kanter 1994, p. 97; italics added). Kanter's conceptualisation of strategic collaboration and alliance mirrors Van de Ven's (1976) conceptualisation of inter-organisational arrangements as living/action social systems which necessitate closer interaction of structural and relational dimensions to make them work. Even though Kanter acknowledges that no two collaborations are the same and do not tread the same path, she nevertheless asserts that "the relationships between companies begin, grow, develop—or fail, in ways similar to relationships between people" (p. 97). In other words although the nuances underpinning each collaboration are specific to it, nevertheless there are commonality in terms of how different inter-organisational collaborations evolve over time. The collaborations develop, according to Kanter (1994), akin to how relationships between human beings develop and terminate, particularly in a marriage. In this context Kanter (1994) identifies five overlapping stages through which "successful collaborations" unfold over time. The five stages include (a) courtship—the phase during which two partners meet and feel attracted towards each other as they discover their respective compatibility; this phase leads to (b) engagement, wherein the partners negotiate, draw up their plans and form contractual relationship; (c) housekeeping is the stage wherein the partners start to work closely with each other and it is in this phase that operational and cultural differences emerge in the collaborations; (d) learning to collaborate, which entails devising mechanisms—structure, process and skills—to resolve differences and to ensure communication, coordination and control; and (e) internal changes.

The focus of each of these studies was on identifying temporal stages or phases that constitute the life cycle of inter-organisational collaborations. The life cycle approach conceptualises the evolution of collaborations as a sequence of predictable life cycle stages wherein each phase logically progresses from or follows the immediate earlier phase. The sequences of life cycle in these studies is essentially linear, unidirectional, irreversible and therefore predictable. The sequences are also cumulative in the sense that characteristics or aspects from earlier stages underpin aspects in the subsequent states. These studies also assume that successful inter-organisational collaborations move smoothly from one phase to the next, and in that context these studies provide a very simplistic and functionalist view of alliance evolution. In doing so, the life cycle approach assumes a mechanistic approach to development of inter-organisational collaborations.

Interestingly, if one explores the policy documents and the White Papers from different UK government sources, making a case for collaboration between the blue light organisations, it is quite clear that there is an underlying assumption that the formation of collaboration would inevitably lead to greater efficiency and effectiveness. Put simply, these emphasise the act of formation of the collaboration per se as central to attain the outcomes (HMIC (2011, 2012); The Knight Report 2013; Policing and Crime Act 2017).

Process Perspective

Within the broad domain of strategy literature, calls for better understanding the processes of change in organisations and organisational forms became popular in early 1990s (Van de Ven 1992; Pettigrew 1992; Van de Ven and Poole 1995). Van de Ven (1992) in his seminal paper titled "Suggestions for Studying Strategy Process" provides an overview of how the term "process" is conceptualised within strategy literature, clarifies what process theory entails and suggests how research could be designed to observe process. He catalogues three dominant definitions and meanings of the term process: "(1) a logic that explains a causal relationship between independent and dependent variables, (2) a category of concepts or variable that refers to the actions of individuals or organisations, and (3) a sequence of events that describes how things change over time" (Van de Ven 1992, p. 169). Scholars who adopt the process perspective essentially focus on identifying and exploring sequence of events that underpin the development of inter-organisational relationships. Ring and Van de Ven (1994) conceptualise evolution of inter-organisational collaboration as "a macro level phenomenon" which nevertheless "emerges, evolves, grows and dissolves over time as a consequence of individual activities" (p. 95). It is the emphasis on the importance of individual activities that underpins their assertion. They further clarify, "cooperative inter-organisational collaborations are socially contrived mechanisms for collective action, which are continually shaped and restructured by actions and symbolic interpretations of parties involved ... just as an initial structure of safeguards establishes a context for interparty action, so also do subsequent interactions reconstruct and embody new governance structures for the relationship" (p. 96). In other words, Ring and Van de Ven conceptualise inter-organisational collaborations as dynamic social systems which are continuously shaped by the actions of and interaction between individuals representing their respective organisations. Initial structures or governance mechanisms arrived at through interaction between the individuals provide conditions for future interaction between them, which further leads to creation of new structures that govern the interaction as well as the relationship as the collaboration develops over time. Based on these conceptualisations, a framework was proposed to explain the developmental processes that underpin the evolution of inter-organisational collaborations. Four key concepts or assumptions underpinned their framework: (a) uncertainties inherent in an inter-organisational collaboration; (b) efficiency and equity criteria are central for assessing the relationship; (c) during the course of the evolution, there would be need for internal resolution of disputes; and (d) the significance of role relationships, which facilitate the development of such relationships. By identifying the four key concepts, they attempt to bring into context the formal, legal and informal social-psychological processes that facilitate the way the organisations jointly negotiate, commit to and execute their relationship as their relationship evolves over time. Ring and Van de Ven posit that the development of alliances consists of "a repetitive *and overlapping* sequence of negotiation, commitment, and execution stages, each of which is assessed in terms of efficiency and equity" (italics added: p. 97). In other words, the evolution of an inter-organisational collaboration is a result of an ongoing interactional process which not only ensures flow of communication but also entails continuous assessment of the relationship based on two criteria—efficiency and equity. The duration of each stage, they claim, varies based on the uncertainty of issues involved, the reliance on trust among the partners of the collaborative venture and the respective role relationships of the partners. Subsequent studies on evolution of strategic partnerships have, in one form or other, expanded Ring and Van de Ven's[3] work.

---

[3] For detail overview of the process literature on inter-organisational collaborations, refer to de Rond (2004) and Patnaik (2011).

Conclusion

Inter-organisational partnerships are a ubiquitous phenomenon and the strategy to form collaboration is considered as the most popular growth strategy. In public sector context in general and in relations to the blue light organisations in particular, strategic partnerships are considered as central to achieve greater efficiency and effectiveness. Notwithstanding the significance of inter-organisational relationships, there is less evidence that, indeed, these partnerships deliver the desired outcomes. Rather, most studies point to the structural and social complexities associated with the phenomenon in explaining the high rate of failure and termination of collaborations. In this respect, one of the major criticisms directed at various theories and perspectives that explain the formation of inter-organisational collaborations is that none of them provides any significant insights on how the collaborations develop over time and how the human agency shapes the developmental process. We argue that adopting process perspective could provide us more meaningful insights on how to manage these complex and dynamic social phenomena. Extant literature on blue light collaborations have made a strong case for the formation of collaborations to attain greater efficiency and effectiveness (see, for instance, HMIC 2012; The Knight Report 2013; The Policing and Crime Act 2017); yet, there is lack of insight on how actually the process of collaboration could be managed. The JESIP review documents (see HM Government Consultation 2016) highlight the challenges the organisations are facing in internalising the interoperability principles, which we argue are actually the relatively 'easier' elements considering these principles are institutionalised structural routines of working together. Therefore notwithstanding the rationale for forming collaborations, other factors such as how collaborations are governed, that is how are the partnerships designed and implemented; how is trust developed and maintained; how knowledge and information sharing takes place; and how tensions and conflicts are managed are critical. These salient aspects sensitise us to the social and cultural aspects that underpin collaborations.

Studies on the dynamic evolution of inter-organisational collaborations tend to focus on the processes of formation, development and dissolution to explain how and why collaborations undergo transformation over time (Ring and Van de Ven 1994; Doz 1996; de Rond and Bouchikhi 2004; Salk 2005). Contractor (2005) notes that by the term 'process', management scholars essentially refer to human interaction and behavioural dimension in organisational and managerial activities.[4] Therefore, studies on the alliance evolution seek to focus on the social relationships that underpin continued interaction between those who organise and manage alliances over time. Hence, scholars who study dynamic process of alliance evolution conceptualise inter-organisational collaborations as a social phenomenon instead of viewing it as a discrete market transaction or as a hybrid organisational arrangement, where cost considerations are paramount. The social nature of inter-organisational collaborations highlights the significance of social relationships and interaction between key individuals involved in collaboration, reciprocity norms, trust and an informal climate underneath the formal structural governance mechanism (Larson 1992; Powell 1998). Simply put, the social aspects underpinning inter-organisational collaboration are intertwined with the governance aspects, in the sense that the governance issues, which are in essence result of social interaction, provide the overarching context for subsequent social interaction in a collaboration. Notwithstanding the call for greater focus on social actors involved in inter-organisational collaborations over time, there is paucity of research that has directly studied processes of change and transformation of collaborations, including in public sector (see, for instance, Vangen et al. 2015). Although it is commonly accepted that collaborations evolve over time and predominantly go through different overlapping phases such as formation implementation and re-evaluation or dissolution phases, the boundaries between the phases are not always clearly demarcated (Das and Teng, 2002; Patnaik, 2011). However what is critical is that each phase in the evolution of inter-organisational collaborations essentially reflects key activities that account for differences in (inter)organisational and managerial behaviour. Therefore, the actions and symbolic interpretations of individual managers and actors shape and recreate the processes of change and transformation (Ring and Van de Ven 1994; Ness 2009). It is against this backdrop that Salk and Shenkar (2001)

---

[4] In this context Contractor also notes that the term 'structure' within alliance research refers to the "legal form of the relationship" (p. 124). However Vangen et al. (2015) have highlighted that in specific context to collaborations in public sector organisations, 'collaborative governance' is used interchangeably. It alludes to the overarching structure that brings together two or more organisations to be governed under one arm. The suggested structure wherein the Police and Crime Commissioner (PCC) becomes responsible for fire and rescue services and the police services is an illustration of the term. Vangen et al. (2015) have differentiated between collaborative governance, which means governance of multiple organisations, from governing collaborations, which allude to how specific collaborations are designed and implemented.

argue that patterns of interactions and actions of individuals reflect the influence of their respective social and cultural identities. Existing studies on individual blue light organisations sensitise us to deeply held cultural values amongst organisational members (see Wankhade et al. 2018; Chan 1997). Therefore, success of collaborations amongst blue light organisations is not simply about why and under what institutional conditions they are formed, rather how each collaboration is managed over time.

REFERENCES

Ansell, C., & Gash, A. (2008). Collaborative governance in theory and practice. *Journal of Public Administration Research and Theory, 18*(4), 543–571.

Barney, J. (1991). Firm resources and sustained competitive advantage. *Journal of Management, 17*(1), 99–120.

Bryson, J. M., Crosby, B. C., & Stone, M. M. (2006). The design and implementation of cross-sector collaborations: Propositions from the literature. *Public Administration Review, 66*(1), 44–55.

Chan, J. B. L. (1997). *Changing police culture.* Cambridge: Cambridge University Press.

Chen, B. (2010). Antecedents or processes? Determinants of perceived effectiveness of interorganizational collaborations for public service delivery. *International Public Management Journal, 13*(4), 381–407.

Contractor, F. J. (2005). Alliance structure and process: Will the two research streams ever meet in alliance research? *European Management Review, 2*, 123–129.

D'Aunno, T., & Zuckerman, H. S. (1987). A life-cycle model of organizational federations: The case of hospitals. *Academy of Management Review, 12*(3), 534–545.

Dacin, M. T., Oliver, C., & Roy, J. (2007). The legitimacy of strategic alliances: an institutional perspective. *Strategic Management Journal, 28*(2), 169–187.

Dacin, M. T., Ventresa, M. J., & Beal, B. D. (1999). The embeddedness of organizations: Dialogue & directions. *Journal of Management, 25*(3), 317–356.

Das, K. T., & Teng, B.-S. (2000). A resource-based theory of strategic alliances. *Journal of Management, 26*(1), 31–61.

Das, T. K., & Teng, B. (2002). The dynamics of alliance conditions in the alliance development process. *Journal of Management Studies, 39*, 725–746.

Davis, G. F., & Cobb, J. (2010). Resource dependence theory: Past and future. *Research in the Sociology of Organizations, 28*, 21–42.

de Rond, M. (2003). *Strategic alliances as social facts: Business, biotechnology and intellectual history.* Cambridge: Cambridge University Press.

de Rond, M., & Bouchikhi, H. (2004). On the dialectics of strategic alliances. *Organization Science, 15*(1), 56–69.

DiMaggio, P., & Powell, W. (1983). The iron cage revisited: Institutional isomorphism and collective rationality in organizational fields. *American Sociological Review, 48*(2), 147–160.

Doz, Y. (1996). The evolution of cooperation in strategic alliances: Initial conditions or learning processes. *Strategic Management Journal, 17*(1), 55–83.

Duysters, G., Heimeriks, K. H., Lokshin, B., Meijer, E., & Sabidussi, A. (2012). Do firms learn to manage alliance portfolio diversity? *European Management Review, 9*, 139–152.

Faulkner, D. O., & de Rond, M. (2000). Perspectives on co-operative strategy. In D. O. Faulkner & M. de Rond (Eds.), *Cooperative strategy: Economic, business, and organizational issues.* Oxford: Oxford University Press.

Forrest, J. E., & Martin, M. J. (1992). Strategic alliances between large and small research intensive organizations: Experiences in the biotechnology industry. *R&D Management, 22*, 41–54.

Franko, L. G. (1971). Joint venture divorce in the multinational company. *The International Executive, 13*, 8–10.

Galaskiewicz, J. (1985). Interorganizational relations. *Annual Review of Sociology, 11*, 281–304.

Greiner, L. (1972). Evolution and revolution as organizations grow. *Harvard Business Review, 50*, 55–67.

Grey, B. (1989). *Collaborating: Finding common ground for multiparty problems.* San Francisco, CA: Jossey-Bass Publishers.

Grimshaw, D., Vincent, S., & Willmott, H. (2002). Going privately: Partnership and outsourcing in UK public services. *Public Administration, 80*(3), 475–502.

Gulati, R. (1998). Alliances and networks. *Strategic Management Journal, 19*(4), 293–317.

Gulati, R., & Gargiulo, M. (1999). Where do interorganizational networks come from? *American Journal of Sociology, 104*(5), 1439–1493.

Harrigan, K. R. (1988). Joint ventures and competitive strategy. *Strategic Management Journal, 9*(2), 141–158.

Hibbert, P., & Huxham, C. (2010). The past in play: Tradition in the structures of collaboration. *Organization Studies, 31*(5), 525–554.

HM Government Consultation Paper. (2015, September). *Enabling closer working between the emergency services.* Retrieved March 20, 2018, from https://assets. publishing.service.gov.uk/government/uploads/system/uploads/attachment_data/file/495371/6.1722_HO_Enabling_Closer_Working_Between_the_Emergency_Services_Consult....pdf\.

HM Government Consultation. (2016). *Enabling closer working between emergency services: Consultation response and next steps.*

HMIC. (2011). *Adapting to austerity.* London: HMIC.

HMIC. (2012). *Increasing efficiency in the police services: The role of collaborations.* London: HMIC.

Huxham, C., & Vangen, S. (2005). *Managing to collaborate: The theory and practice of collaborative advantage Abingdon.* London: Routledge.

Inkpen, A. C., & Tsang, E. W. K. (2007). Learning and strategic alliances. *Academy of Management Annals, 1*(1), 479–511.

JESIP. (2016). *The Joint Doctrine: The interoperability framework.*

Kale, P., & Singh, H. (2009). Managing strategic alliances: What do we know now, and where do we go from here? *Academy of Management Perspectives, 23*(3), 45–62.

Kanter, R. M. (1994). Collaborative advantage: The art of alliances. *Harvard Business Review, 72*(4), 96–108.

Knight, K. (2013). *Facing the future: Findings from the review of efficiencies and operations in fire and rescue authorities in England.* London: HSMO.

Larson, A. (1992). Network dyads in entrepreneurial settings: A study of the governance of exchange relationships. *Administrative Science Quarterly, 37*, 76–104.

Lunnan, R., & Haugland, S. A. (2008). Predicting and measuring alliance performance: A multidimensional analysis. *Strategic Management Journal, 29*(5), 545–556.

McGuire, M. (2006). Collaborative public management: Assessing what we know and how it. *Public Administration Review, 66*(1), 33–43.

Murray, A. E., & Mahon, J. (1993). Strategic alliances: Gateway to the New Europe? *Long Range Planning, 26*, 102–111.

Ness, H. (2009). Governance, negotiations, and alliance dynamics: Explaining the evolution of relational practice. *Journal of Management Studies, 46*(3), 451–480.

Niesten, E., & Jolink, A. (2015). Alliance management capabilities and performance. *International Journal of Management Reviews, 17*, 69–100.

O'Leary, R., & Bingham, L. (2007). Conclusion: Conflict and collaboration in networks. *International Public Management Journal, 10*(1), 103–109.

Oliver, C. (1990). Determinants of interorganizational relationships: Integration and future directions. *The Academy of Management Review, 15*(2), 241–265.

Oliver, C. (1996). The institutional embeddedness of economic activity. *Advances in Strategic Management, 13*, 163–186.

Osborne, S. P. (Ed.). (2009). *The new public governance? New perspectives on the theory and practice of public governance.* London: Routledge.

Parkhe, A. (1993). "Messy" research, methodological predispositions, and theory development in international joint ventures. *Academy of Academy of Management Review, 18*(2), 227–268.

Patnaik, S. (2011). *Inter-organisational collaborations as embedded social systems: A critical realist explanation of alliance evolution.* Unpublished Doctoral Thesis, University of Liverpool.

Pettigrew, A. M. (1992). The character and significance of strategy process research. *Strategic Management Journal, 13*, 5–16.

Pfeffer, J., & Salancik, G. (1978). *The external control of organizations a resource dependence perspective.* New York: Harper & Row.

Powell, W. W. (1998). Learning from collaboration: Knowledge and networks in the biotechnology and pharmaceutical industries. *California Management Review, 40*(3), 228–240.

Provan, K. G., & Kenis, P. (2008). Modes of network governance: Structure, management, and effectiveness. *Journal of Public Administration Research and Theory, 18*(2), 229–252.

Provan, K. G., Keith, G., Fish, A., & Sydow, J. (2007). Interorganizational networks at the network level: A review of the empirical literature on whole networks. *Journal of Management, 33*(3), 479–516.

Provan, K. G., & Milward, H. B. (2001). Do networks really work? A framework for evaluating public-sector organizational networks. *Public Administration Review, 61*(4), 414–423.

Reuer, J. J., Zollo, M., & Singh, H. (2002). Post-formation dynamics in strategic alliances. *Strategic Management Journal, 23*, 135–151.

Ring, P. S., & Van de Ven, A. H. (1994). Developmental processes of cooperative inter-organizational relationships. *Academy of Management Review, 19*(1), 90–118.

Salk, J. E. (2005). Often called for but rarely chosen: Alliance research that directly studies process. *European Management Review, 2*(2), 117–122.

Salk, J., & Shenkar, O. (2001). Social identities in an international joint venture: An exploratory case study. *Organization Science, 12*(2), 161–178.

Scott, W. R. (1995). *Institutions and organizations.* Thousand Oaks, CA: Sage.

Spekman, R. E., Forbes, T. M., Isabella, L. A., & MacAvoy, T. C. (1998). Alliance management: A view from the past and a look into the future. *Journal of Management Studies, 35*(6), 747–772.

The Policing and Crime Act. (2017). *UK Parliament.*

Thomson, A. M., & Perry, J. L. (2006). Collaboration processes: Inside the black box. *Public Administration Review, 66*(1), 20–32.

Van de Ven, A. (1976). On the nature, formation and maintenance of relations among organizations. *Academy of Management Review, 1*(4), 26–37.

Van de Ven, A. H. (1992). Suggestions for studying strategy process: A research note. *Strategic Management Journal, 13*, 169–188.

Van de Ven, A., & Poole, M. (1995). Explaining development and change in organizations. *Academy of Management Review, 20*(3), 510–540.

Vangen, S., Hayes, J. P., & Cornforth, C. (2015). Governing cross-sector interorganizational collaborations. *Public Management Review, 17*(9), 1237–1260.

Wankhade, P., Heath, G., & Radcliffe, J. (2018). Cultural change and perpetuation in organisations: Evidence from an English emergency ambulance service. *Public Management Review, 20*(6), 923–948.

Wood, D. J., & Gray, B. (1991). Toward a comprehensive theory of collaboration. *The Journal of Applied Behavioral Science, 27*(2), 139–162.

Yan, A., & Zeng, M. (1999). International JV instability: A critique of previous research, a reconceptualization, and directions for future research. *International Journal of Business Studies, 30*, 397–414.

Zaheer, A., & Harris, J. (2006). Inter-organizational trust. In O. Shenkar & J. Reuer (Eds.), *Handbook of strategic alliances.* Thousand Oaks, CA: Sage.

CHAPTER 3

Building Strategic Capacity and Collaborative Leadership in Blue Light Organisations

Abstract It is increasingly considered that an organisation's ability to form and manage strategic partnerships significantly contributes in enhancing its overall performance. Coordination, communication and ability to develop interpersonal relationships (bonding) are considered as three critical components of collaborative capabilities. The collaborative capabilities develop over a period of time, and they enable the organisation to purposefully create, extend or modify existing organisational routines that underpin the activities pertaining to coordination, communication and relationship building. Development of collaborative capabilities necessitates exploring alternative approaches to leadership in organisations.

Emergency services leadership has been characterised as 'top-down', hierarchical, 'heroic', with a command and control approach prevalent in the organisations. There has been reliance on historical and hierarchical models of 'heroic' and 'top-down' leadership and absence of a distributive and pluralist approach to leadership. Current thinking and models are often based around individual services without much joined-up approach. Greater collaboration entails an approach different from leadership development, which needs to be facilitated at multiple levels within the organisations. Development of collaborative culture in organisations will necessarily involve cultivating future leaders, who will encourage greater collaboration within and amongst the collaborating organisations.

© The Author(s) 2020
P. Wankhade, S. Patnaik, *Collaboration and Governance in the
Emergency Services*,
https://doi.org/10.1007/978-3-030-21329-9_3

Keywords Collaborative capabilities • Leadership theories • Collaborative leadership • Capacity development

INTRODUCTION

Strategic partnerships are characterised by three distinctive overlapping features. First, these arrangements are characterised by greater uncertainty and ambiguity. Second, the partners do not have specific insights at the beginning of the partnership on how they could create greater partnership together than they would on their own. Value, in this context, does not merely have a financial or economic connotation and rather encompasses a broader meaning, which also emphasises quality of service, as is the case with the blue light organisations. Third, how collaborations would evolve over time is hard to predict, at least at the beginning of the partnership. Apart from these distinctive facets, which are generic to almost all collaborations, partnerships between blue light services, more than private organisations, are conceptualised and implemented as organisational change initiatives. It is in this respect that collaborative leadership and capacity building assume further significance. Collaborative leadership, we argue, entails providing leadership to guide and manage two distinct processes, that is, the strategic partnership which evolves through different overlapping phases, including alliance formation, implementation and re-evaluation, and dissolution (Ring and Van de Ven 1994; Doz 1996; Das and Teng 2000); and the change process that primarily involves the phases of unfreezing, changing and refreezing (Lewin 1947). Greater involvement of organisational members or employees is central to achieve successful integration of the partner organisations.

Organisational capacity is generally referred to an organisation's ability to perform its overarching activities (Yu-Lee 2002) and it entails identifying and enhancing factors that significantly contribute in facilitating an organisation to perform its activities to attain its goals and objectives. Cox et al. (2018) have provided a broad differentiation on how the concept, organisational capacity, is conceptualised in public and private sector organisations. Whereas in the public sector, organisational capacity is conceptualised as the ability of the government to allocate, develop, direct and control resources, namely, human, financial, physical and information resources, in contrast, in the private sector, the concept is often conceptualised as a set of processes, practices and attributes that contribute in facilitating an organisation achieve its overarching objectives. In both sets of interpretation, organisational capacity is considered as critical in influencing performance and effectiveness of an organisation.

The chapter is as structured. First, considering our interest on collaborations and strategic partnerships, we focus our attention on alliance or collaborative capabilities, and in that context we explore the nuances underpinning collaborative leadership. Broadly speaking collaborative capabilities highlight the capacity of an organisation to effectively form and manage partnerships, whereas collaborative leadership highlights the role that organisational and divisional/departmental leaders can play in (a) actively influencing outcome of specific partnerships and (b) creating organisational culture that could contribute in the development of collaborative capabilities. Then we pay attention to the notion of collaborative leadership and explore some of the key leadership theories and approaches. We conclude the chapter by reflecting on the implications of leadership in collaborative settings.

COLLABORATIVE CAPABILITIES

In the gamut of research on inter-organisational collaborations, alliance or collaborative capabilities is a recent construct that has captured attention of academic scholars and policymakers. It is conceptualised as an organisational-specific capability that enhances an organisation's capacity or ability to form, organise and manage alliances, and the development of such capabilities entail efforts to effectively capture, share and disseminate collaboration-specific know-how associated with prior experiences (Kale et al. 2002; Kale and Singh 2007; Schreiner et al. 2009). Primarily scholars studying such organisational capabilities draw on theoretical insights from organisational learning and dynamic capabilities literature and posit that organisations develop collaborative capability by accumulating, integrating and internalising knowledge and skills gained by participating in different alliances with various partners and modifying these as situations evolve (Simonin 1997; Anand and Khanna 2000; Heimeriks and Duysters 2007; Gulati et al. 2012). This perspective is also consistent with the conceptualisation of organisational capabilities as embedded in activities and routines within organisations for addressing often complex, practical and repeated problems (Schreyögg and Kliesch-Eberl 2007). Recent studies suggest that organisations possessing such capabilities attain greater success in their collaborative activities (Dyer and Singh 1998; Gulati 1998;

Anand and Khanna 2000; Ireland et al. 2002; Kale and Singh 2007; Heimeriks and Duysters 2007; Schreiner et al. 2009; Arikan and McGahan 2010). It is important to highlight here that notwithstanding increasing collaborative activities performed by public sector organisations, including blue light organisations, there remains a gap in existing knowledge on (a) how such organisations manage their collaborations and (b) what they learn and internalise from collaborative activities.

CONSTITUENT ELEMENTS OF COLLABORATIVE CAPABILITIES

Previous studies have highlighted three broad managerial challenges associated with inter-organisational relationships, including those between blue light organisations. First, strategic partnerships result in greater interdependency and divided authority structure and the cognitive and cultural distance, in particular, between partners create conditions that lead to coordination lapses. Therefore, it is of essence, that appropriate interfaces and boundary-spanning mechanisms are created to operationalise collaborations (Gulati et al. 2012). Hence Luo (2006) suggests that organisations must create adaptable mechanisms in terms of formal and routine procedures, rules and policies to guide cooperation between partners and create an appropriate framework for their ongoing interaction. These mechanisms, it is assumed, would facilitate smoother coordination of activities leading and contributing to addressing the challenges pertaining to divided authority structure. The second management challenge in strategic partnerships pertains to lack of information sharing and communication, which results from dearth of understanding of respective partners' structural and cultural idiosyncrasies, as well as limited attempts to build shared understanding of the partners' obligations and more importantly develop shared mental models of how to work together effectively (Mohr and Spekman 1994; Schreiner et al. 2009; Gulati et al. 2012). In specific context to functioning and coordination of emergency services, The Kerslake Report (2017), following the Manchester Arena terrorist attack as well as the initial insights from the ongoing Grenfell Tower fire enquiry, points to coordinational and communicational failure between the emergency services in responding to the complexities arising from these incidents. In his report, Lord Kerslake specifically referred to the collaboration between emergency services and other responder agencies to enhance the multi-agency command, control and coordination of response to major incidents; this has led to the development and adoption of the JESIP

Interoperability Framework. The JESIP Interoperability Framework is underpinned by five key concepts, which include co-location, communication, coordination, joint understanding of risk and shared situational awareness; these are considered to enable responders to work together effectively. These five concepts, which in essence, seen through the lens of collaborative capabilities, could be organised under the categories of 'coordinational' and 'communicational' challenges. Interestingly, in his report, Lord Kerslake (2017) categorically highlighted that the coordinational and communicational failures amongst the emergency services and responders took place in spite of the existence of the JESIP Interoperability Framework. He specifically pointed out that the failure of coordination and communication was, in essence, due to lack of implementation of formal procedures as stated in Her Majesty's Government guidance (2013). The findings from the tri-service review of the JESIP (2016) that explored the embeddedness of JESIP across emergency services highlight that the principles underpinning the joint decision-making model were neither widely accepted nor understood (p. 14/15) and this was particularly so in the case of members who were located in lower levels of hierarchy. Similarly, the members at senior and middle management level have had some form of JESIP training as compared to members belonging to operational level, who also did not consider that the training they were provided with has prepared them to work effectively across services (p. 20/21). Lord Kerslake's Report, as well as the insights from the review of JESIP embeddedness in emergency services (2016), highlights the significance of the establishment of shared coordinational and communicational routines and processes that could facilitate interoperability across the blue light services in the events of a crisis. Lord Kerslake's Report (2017) and the JESIP embeddedness review (2016) also raise a critical point, that is, existence of formal structures and procedures do not necessarily lead to effective response management unless the partner organisations have invested time and resources to formalise routines and practices to reduce pervasive uncertainty in decision-making (Becker and Knudsen 2005; Feldman and Pentland 2003), which is particularly distinctive in regards to what blue light organisations actually do in crisis situations.

It is clear that culturally, emergency service organisations are distinctly different from one another. Davis and Thomas (2003), in their critique of police culture, which has also been an area of considerable interest for many scholars since the 1960s and 1970s, conclude that it is indeed rigid, immutable and "deeply pervasive" (p. 682), which emphasise crime fighting and in the process legitimises masculine subjectivity (also see the works of Reiner 2010; Loftus 2010). Some of the cultural attributes that Bullock et al. (2006) note were extended to the concept of partnership working which was introduced in the 1990s. However, O'Neill and McCarthy (2014) in their more recent work on partnership between police forces and other community agencies, particularly in the context of neighbourhood policing, found that police officers involved in partnerships considered them to be effective and crucial to their work and in some cases found working in such settings more enjoyable. Chapman (2015), in her study that involved a group of police officers and ambulance staff, observed that members from both the professions held each other in high regard that was underpinned by "trust, professional respect, good rapport and a mutual understanding of the roles that each perform" (p. 163). Chapman's (2015) work concludes with an interesting insight that although police officers and ambulance staff demonstrated better understanding of each other's cultural characteristics and viewed each other's roles and work in high regards, they did not hold the same view towards colleagues from the fire services. Possessing and developing positive view of each other's work and expertise is the starting point of what alliance scholars identify as the third managerial challenge that inhibits strategic partnerships. Developing interpersonal relationships is considered as central to effective functioning of collaborations (Ring and Van de Ven 1994; Spekman et al. 1996; Arino et al. 2005; Gulati et al. 2012). Close personal relationships and bonding among individuals create conditions for establishing norms of trust which serve as a conduit for creating and maintaining expectations of mutual cooperation as well as facilitating knowledge sharing between organisational members. Interpersonal relationships are also critical in mitigating conflict situations as they arise. In other words, bonding matters, and in that context, having a positive view of each other's work and roles creates conditions for development of interpersonal rapport between organisational members. In this respect, Chapman's (2015, p. 171) assertion that "when seeking strategies for enhanced interoperability between organisations, due consideration should be given to the advantages and benefits of human interoperability, which are sometimes neglected at the expense of hardware solutions" rings true. Creating a culture that values learning from each other, particularly amongst organisations working in close proximity with each other, as is the case with blue light organisations, should rank high in the list of priorities for leaders of the three services.

COLLABORATIVE LEADERSHIP

Interestingly, although collaborations are recognised as a distinctive feature of public administration in general and increasingly in the context of emergency services (see Wilson and Grammich 2017), the literature on leadership in public management does not adequately engage with nuances underpinning leadership in collaborative contexts (Huxham and Vangen 2005; Crosby and Bryson 2010; Morse 2010). This is not simply a gap in the literature in public management; rather, this is also observed in the extant literature on inter-organisational relationships. In both sets of literature, scholars tend to use the terms 'managers', 'senior management', 'boundary agents' and 'alliance champions' interchangeably to often highlight the role of the social agents involved in the formation and operationalisation of strategic partnerships (Crosby and Bryson 2010; Kale and Singh 2009; Morse 2008). Notwithstanding this acknowledgement of the role of social agents in shaping, influencing and managing strategic partnerships, there is an increasing call to better explore issues pertaining to leadership for collaborations (Sullivan et al. 2012; Crosby and Bryson 2018).

LEADERSHIP THEORIES AND APPROACHES

The topic of leadership is one of the central concepts in the broad domain of public management or public administration and it has attracted considerable attention from different scholars, particularly in the context of its significance and challenges associated with leadership issues (Van Wart 2003; Getha-Taylor et al. 2011). In a recent critique on leadership in public management, Chapman et al. (2016) conclude that "while leadership in general has been difficult to study and measure, the increasing complex scope of public administration has made this more challenging, resulting in a fragmented approach to the study of public service leadership" (p. 111). Collaborations and various types of strategic partnerships have added to the existing complexities associated with leadership in public management (Page 2010; Crosby and Bryson 2018). Broadly speaking, there are three strands of research on leadership in the context of public sector management. One body of research focuses on the character underpinning public leadership, whereas the second stream of studies explores functions associated with leadership and which include the importance of accountability, strategic actions, collaborations and entrepreneurial initiatives. The third stream of research highlights the jurisdiction of leadership in public management, that is, where leadership is exercised (Getha-Taylor et al. 2011; Van Wart 2013). Leadership issues in collaborative arrangements, in essence, occupy space within each of the three strands.

Most reviews of leadership theories begin with a discussion and critique of trait leadership approach. The trait leadership theories, which became popular in the 1920s and 1930s (see Van Wart 2003), conceptualised leadership in terms of traits and posit the argument that individuals were born leaders. Hence, the leadership studies during this period focused on the identification of "magic personality traits" (Fiedler 1997, p. 126) and other distinctive traits that are associated with the "great man" (Van Wart 2003, p. 2). In an in-depth review of leadership theories, Bass (1990a, b) notes that scholars who explored leadership traits, essentially examined leader's traits in relation to different aspects, such as demographic (e.g. gender, age, ethnicity, social status and education), task competence (e.g. intelligence, conscientiousness, openness to experience, emotional stability, technical knowledge and leadership self-efficacy) or interpersonal attributes (e.g. agreeableness, extraversion, communication skills, emotional intelligence and political skills). The trait perspective posits the view that traits are relatively fixed and in fact represent an enduring aspect of a leader's personality and therefore traits could facilitate identification of potential leaders, thus making it feasible to purposely select individuals possessing those traits. Trait leadership approach has been criticised, for example, for lack of agreement about essential leadership traits, lack of clarity in terms of distinctions amongst the traits, failure of proponents to the context in which leadership manifests and lack of utility with regards to training and development considering the permanence of traits (Northouse 1997; Derue et al. 2011).

Notwithstanding these criticisms, Drodge and Murphy (2002) highlight that trait leadership approach is still widely referred to in the context of police leadership. They note that "the issue of police leadership commonly comes under scrutiny whenever police actions are deemed controversial, or when there is a dramatic shift in organisational structure and philosophy … in both the cases, it is often leadership trait that get highlighted by the public, through the media, and by the rank—and—file police officers themselves" (Drodge and Murphy 2002, p. 203). This aspect, in a sense, supports the assertion that leadership in police force organisations still reflects the 'quasi-military' structure and continued focus of researchers on individual leaders and their skills reinforces this viewpoint (Herrington and Colvin 2016; also see Davis and Bailey 2018).

In the 1940s the focus of leadership research shifted from identifying traits and attributes to conceptualising leadership in terms of their behaviour. The central argument underpinning the behavioural theory is that leadership behaviours reflect either a 'task orientation', which is a concern for production and achievement of objectives, or a 'relations orientation', which is a concern for the needs and interests of followers. The task-orientation approach is underpinned by efforts of the leaders to define task roles and role relationships, coordinating actions of organisational members, determining standards of task performance and developing reward mechanism. Simply put, creation of structures and routines is considered as critical to shape behaviour, commitment and motivation of followers or subordinates. In contrast, relational-oriented behaviour of leadership emphasises concern and respect for individual group members and depicts leaders as friendly and approachable, open to input from their colleagues and subordinates and who treat organisational members as equals (Bass 1990a, b). The central argument underpinning this perspective is that leaders act in ways that demonstrate concerns for the needs and interests of the followers and, in the process, it contributes in building respect amongst followers.

Both trait leadership and behaviour leadership approaches were criticised for overlooking the social context in which leadership manifests (Van Maurik 2001). Situational leadership theorists tried to address this anomaly by first arguing that there was no universal 'best' way to influence people, but, rather, successful leadership, in essence, entails adapting one's leadership style to the demands of the circumstances or situation. Hersey et al. (2008) conceptualised leadership style as a pattern of task behaviours and relationship behaviours, reflective of task and relations-oriented classifications from earlier theories. Task behaviour is focused on ensuring tasks are completed compared with relationship behaviour which prioritises relationships and support of individuals. In this backdrop, Hersey et al. (2008) identified four leadership styles. 'Telling' (or directing) emphasises a directive approach with a focus on task rather than the relationship and support of the individual. This is particularly when a follower is unable and insecure to perform the task on his or her own, and hence needs guidance and directives. The 'selling' (or coaching) style is directed at followers who are unable to perform tasks though they show willingness and desire to act. In this context, the role of the leader is to be both directive as well as supportive. The fundamental difference from telling is that the leader is open for dialogue and clarification in such a way that the follower "buys" the idea, internalises it and feels ownership to it. The guidance and emotional support from the leaders, it was suggested, would uphold the motivation of the followers. 'Participating' (or supporting) is the third leadership style, and it includes both directive and supportive elements. At one level it highlights high level of trust in the abilities of the follower and yet at the same time the leader acknowledges that the follower would require more social support and recognition, and fewer directions to perform the tasks. This style is characterised by a two-way communication or consultation. Finally, 'delegating' (observing) is the most passive leadership style of the four, as it essentially entails followers who are able and willing to perform the tasks and hence leaders must encourage such followers to take initiatives and perform the tasks independently. It is critical that leaders encourage followers to take initiative, who can handle the planning, performing and control of their tasks autonomously.

Bass (1998) proposed that, broadly, there are two leadership philosophies: transactional leadership and transformational leadership. A transactional leader focuses on performance and attainment of specific tasks and for that incentivises the subordinates with rewards. Failure or poor performance of task results in sanctions. Put simply, the emphasis of a transactional leader is ensuring that tasks are performed within time and scope and following basis training and monitoring. The overarching emphasis of trait and behaviour leadership approaches corresponds to the transactional leadership philosophy (Bass and Bass, 2009; Bass et al. 1996). Transformational leadership, in contrast, emphasises that leaders must thrive to transform the basic values, beliefs and attitudes of followers to motivate them to perform at a higher level. In this context, transformational leadership encompasses not just the leaders but also the followers or subordinates as the intrinsic elements of the leadership process (Van Wart 2013; Drodge and Murphy 2002; Bass 1990a, b). It is critical to highlight here that transformational leadership was first developed by Downton (1973) and later popularised in leadership literature by James Burns (1978). Burns was influenced by insights from organisational psychology, particularly from Maslow's (1954) seminal work on 'hierarchy of needs'. He argued that transformational leadership is linked to psychological fulfilment and it must aim to address people's higher-order needs for achievement, self-esteem and self-actualisation.

Bass (1985) developed these ideas further but emphasised that transactional and transformational leadership must be conceptualised as existing as a continuum, that is, both the leadership approaches are inseparable and not distinct or independent of one another. Bass (1990a, b) conceptualised transformational leadership as comprising of seven factors that combined transformational and transactional approaches with 'laissez-faire' leadership. It classified transactional leadership as contingent reward and management-by-exception and included four transformational factors, also referred to as 'The Four I's', namely, individualised consideration, idealised influence, inspirational motivation and intellectual stimulations. Individualised consideration refers to the level of support that organisational members feel from their formal leaders. The quality of such support positively affects commitment of organisational members. The idealised influence pertains to fostering of trust and respect in the relationship between members of the organisation and the leaders. Idealised influence relates to the issue of ethics and values of the leaders and the followers. Put simply, idealised influence brings forth the notion of ethical thinking and actions. Inspirational motivation relates to the capacity of the leaders to motivate organisational members to achieve more than what they had originally expected. Such leadership, according to Drodge and Murphy (2002), entails presenting a vision that unites people around a common, desirable and tangible goal. Intellectual stimulation, the fourth dimension of transformational leadership, involves encouraging subordinates to challenge existing paradigms and practices and exploring new ways of addressing existing and new challenges (Kelloway and Barling 2000). The critical point to highlight here is that the transformational leadership theory conceptualises leadership as a social process that involves leaders occupying formal positions, organisational members and the context in which leadership is manifested. Put simply, transformational leadership is a process of changing how subordinates feel about themselves, which in turn raises their motivation and enables them to achieve higher level of performance.

One could argue that the social context in which leadership manifests underpins the difference between transactional and transformational leaderships. In this respect, Shivers-Blackwell (2006) notes that, "while an ordinary transactional leader's role, also defined as a 'mechanistic' leader, is merely to be more effective in predictable and more stable environments, a transformational leader, also characterized as an 'organic' leader, will be more effective for organizations operating in unpredictable or even hostile environments" (p. 29). The popularity of transformational leadership since 1980s has coincided with an increasing drive for implementation of reforms in public sector organisations. This is particularly so in the case of leadership in emergency service organisations. Drodge and Murphy (2002), for instance, attribute the influence of transformational leadership in Royal Canadian Mounted Police (R-CMP) to successful development and adoption of community policing and to the creation of the Community, Contract and Aboriginal Policing Services Directorate. They assert, "policies, procedures and new directorates by themselves are not responsible for community policing, rather its accrued success are a measure of some degree of transformational leadership that fostered the sense of support, stability and subtle movement towards goals" (Drodge and Murphy 2002, p. 209). Although research on leadership styles in fire services context is underdeveloped, there have been calls to adopt transformational approach to leadership in the services. In this respect, Alyn (2010) collated survey response from more than 1000 fire service personnel in the USA and concluded that transformational leadership had direct correlation with commitment of organisational members. These support the findings of Pillai and Williams (2004), who also linked transformational leadership approach to higher group cohesiveness, commitment and performance of fire service personnel. Likewise, evidence from the UK Ambulance services also points to greater drive towards need for and adoption of transformative leadership to facilitate organisational change. The Future Leaders Study (2009), which was commissioned by the Ambulance Trusts in England following the comprehensive review of the ambulance services in 2009 by Peter Bradley, categorically makes the case for transition from transactional approach to transformational leadership. The study states:

> Throughout the Future Leaders study, we heard consistent reference for ambulance leaders to shift the prevailing culture from one where the default leadership style is transactional (a directive, 'command and control style') to one that is transformational in nature ... the Future Leaders benchmark tool was an attempt to combine elements of transformational leadership—the qualities, skills, knowledge and behaviors that senior leaders will require to lead ambulance services and organisations—informed by the context in which these characteristics will be applied. (pp. 12–13)

The study concludes that, as organisations, Ambulance Trusts, lacked individuals who could handle and manage the transition towards transformative leadership.

The study found that, as an organisation, Ambulance Trusts lacked individuals who could handle change.

In the specific context to emergency services, there has been greater calls to go beyond transformational services to embrace shared or distributed leadership approach. Proponents of shared or participatory leadership critique transactional and transformational leadership approaches on two grounds. First, they argue that in both the cases, power resides with the leaders in the sense that only the leaders have the power to influence others and the process of influencing also reflects and reinforces the display of power (Yulk 2010; Collinson 2011). Put simply, power is located within individuals and hence is viewed as an attribute or property of leaders. The second criticism is directed at the two approaches overlooking the changing nature of work environment, particularly the complex and collaborative context which underpins contemporary organisational setting. Hence, the suggestion that horizontal approach to leadership could be more useful.

Van Wart (2013) traces the origin of the concept of distributed (horizontal) leadership to 1970s, whereas Gronn (2008) and Leithwood et al. (2009) trace the notion of distributed leadership or 'diffusion of leadership' to the works of Benne and Sheats (1948) and Gibb (1954). However, there is no single definition or conceptualisation of distributed leadership. According to Bolden et al. (2008), distributed leadership "argues for a less formalized model of leadership where leadership responsibilities are disassociated from the organisational hierarchy. It is proposed that individuals at all levels in the organisation and in all roles can exert leadership influence over their colleagues and thus influence the overall direction of the organisation" (Bolden et al. 2008, p. 11). Interestingly, this definition emphasises the notion of influence but neglects the concept of power. Nonetheless a significant number of recent studies have tried to capture application of distributed or shared leadership in organisations related to the education (Devos et al. 2014; Heikka 2014; Holt et al. 2014) and healthcare sectors (West et al. 2014). The UK's Department of Health developed a Clinical Leadership Competency Framework (CLCF) (2011) and called for embracing shared leadership in clinical setting. It states:

> applying to all engaged in clinical practice the CLCF is built on the concept of shared leadership where leadership is not restricted to people who hold designated leadership roles, and where there is a shared sense of responsibility for the success of the organisation and its services. Acts of leadership can come from anyone in the organisation, as appropriate at different times, and are focused on the achievement of the group rather than of an individual. (Department of Health 2011, p. 5)

The central idea that underpinned this argument was that in healthcare, tasks are more complex and highly interdependent, and although all clinicians cannot be expected to be leaders, they can contribute to the leadership process. In this respect, the CLCF identified five core domains of how the clinicians could contribute to shared leadership. The domains are (a) demonstrating personal qualities, (b) working with others, (c) managing services, (d) improving services and (e) setting direction.

Building on the insights from CLCF, Taylor and Armitage (2012) made the case for shared leadership in ambulance services. Specifically, they argued that like clinicians, the work of the paramedics is also complex and that paramedics and ambulance crew rely on dispersed knowledge and expertise in performing the tasks to the highest quality of service. They call for specific training and development interventions "that could enable paramedics to connect with 'leadership' in a very practical way and support an increased awareness and recognition of the importance of clinical leadership as an integral part of day to day clinical practice" (Taylor and Armitage 2012, p. 565). In this regard, they suggested that two approaches, namely the Crew Resource Management (CRM) and simulation, which are widely used in the airlines industry, could be of immense value. Interestingly, the calls for shared or distributed leadership also find support from the UK's College of Policing. The notion of shared leadership underpinned the development of the nine 'guiding principles for organisational leadership' (College of Policing 2017). The guiding principles were developed following the Leadership Review that the College of Policing published in 2015. The guiding principles were organised under three broad themes, namely (a) understanding leadership, (b) displaying leadership and (c) developing leadership. In specific context to 'displaying leadership', the report notes that, "senior leaders should consider team makeup when deciding how an organisation is structured, including multi-agency teams … diversity and differences within teams must be valued and related to complementary skillsets, different professional backgrounds, leadership styles and protected characteristics" (College of Policing 2017, p. 18). In other words, distributed leadership helps in the participation of organisational members, who might belong to different social and demographic sections of the society and who might possess complementary skill sets in resolving complex problems. Notwithstanding these developments, the notion of shared and distributed leadership does not completely address the criticisms associated with transactional and transformational leadership approaches. Distributed leadership, in making the case for use of diverse knowledge and expertise residing within organisational boundaries, essentially sidesteps the issues relating to power. There is an assumption that in the process of fostering greater engagement, the nuances pertaining to power and control could be addressed. Van Wart (2013), in his review of public service leadership, concurs with this view point and asserts that some degree of horizontal leadership is necessary as soon as organisations start sharing power through partnerships and collaboration initiatives.

The notion of power, in essence, underpins leadership in interorganisational settings or collaborative arrangements (Van Wart 2013). 'Collaborative advantage', a term popularised by Huxham and Vangen (2005), makes the case for potential benefits of partnerships across organisational boundaries to achieve results that could not otherwise be achieved by any single organisation. Considering that collaborations entail the notion of working together to attain common goals, it is imperative that leaders in the collaborating organisations would cede and share some of their 'powers' to make the relationship work (Crosby and Bryson 2005). Simply put, collaborative leadership "deemphasizes the roles of both leaders and followers in order to emphasize the needs of the network, system, environment, or community, resulting in a collaborative style" (Van Wart 2013, p. 559). Interestingly Morse (2010) and other scholars including Crosby and Bryson (2010) argue that better understanding of collaborative leadership in the public service context needs greater exploration of their very nature and in this respect they suggest that some of the collaborations and partnerships have greater integrative, rather than mere partnership nature. Morse (2010, p. 232) clarifies the term integration as "to integrate means to bring together and combine or incorporate different components into a whole ... integration, in its ideal form is more than cooperating in order to meet one's own ends. It represents a whole that is greater than the sum of its parts." Seen through this lens, one could argue that the collaboration between the blue light emergency services has a more of an integrative rather than a collaborative orientation. The overarching objectives of blue light collaborations are to attain superior efficiency or effectiveness by functioning under one umbrella (Policing and Crime Act 2017). Hence, it is the statutory duty of the Police and Crime Commissioners (PCCs) to keep collaboration opportunities under review and to collaborate where it is in the interest of their efficiency or effectiveness.

The conceptualisation of inter-organisational relationships as integration brings to forth different set of dimensions as far as the role of leadership in such arrangements is concerned. Morse (2010) posits that leadership in integrative settings is akin to "catalysis", in the sense that leaders act as "catalysts" in accelerating or de-accelerating the process of integration. In this respect Morse (2010) and other proponents of integrative perspective concur with assertions of Huxham and Vangen (2005) that mainstream leadership theories do not translate well in a collaborative context, where leadership is enacted through three interconnected media—structures (formal and informal), processes (formal and informal instruments) and participants (actors). The leaders use different tactics, including farming and setting agenda, convening stakeholders and structuring deliberations in driving the process of integration (Page 2010). The integrative leadership perspective acknowledges that strategic partnerships could create differences and conflict situations with various stakeholders and, in that respect, the leaders use different media and tactics to drive their agenda. This approach of driving the integration process is influenced by Huxham and Vangen's (2005) approach to leadership that posits that leadership in collaborations is essentially about looking for "what makes things happen in a collaboration" (p. 202). Put simply, considering collaborative arrangements between different organisations presents opportunities for synergies, and to create value, it is imperative that leaders in these organisations can create collaborative advantage by actively synthesising the differences (Vangen 2017). Although integrative leadership adds to the existing debates on how to organise and manage inter-organisational relationships in the broad domain of public services, it presents a partial and restrictive picture of leadership. In making the case for fostering integration, it rightly assumes that the relationship between the organisations in such partnerships could be fraught with tensions and conflicts (Das and Teng 2000; de Rond and Bouchikhi 2004) and organisational leaders, using different media and tactics, must drive the process. However, this perspective masks the tensions within the partner organisations that characterise inter-organisational partnerships (see, for instance, Liu et al. 2017; Cartwright and Cooper 1990). In this respect, it is critical to highlight here that the Leadership Review (2015) and the Guiding Principles for Organisational Leadership (2017) for police forces emphasise that the morale and well-being of organisational members are distinctive aspects of leadership. Interestingly, seen strictly through the lens of power relationships, the difference between integrative leadership and other leadership theories is minimal in the sense that the leadership is still conceptualised as the capacity of the individual or groups to influence others, albeit in a more complex organisational and institutional setting.

Over the last decade, all the three emergency services have undertaken respective reviews (College of Policing 2015, 2017; NHS Ambulance Chief Executive Group 2009) to identify the evolving operating environment and leadership qualities needed to drive the changes. All the reviews have concluded that the respective organisations must embrace shared or distributed form of leadership. However, more recent research shows that in spite of significant emphasis on shared leadership, the "notion of authority of rank" (Davis and Bailey 2018), demonstrating the command and control structure, remains deeply embedded. Added to the challenges of embracing alternative leadership approaches (also see Herrington and Colvin 2016), the success and effectiveness of leadership is increasingly conceptualised in managerial terms, which also underpin the underlying discourse in the various reviews (see Wright 2000; Butterfield et al. 2005). Capacity to collaborate is also considered a distinctive leadership quality. For instance, the Future Leaders Study (2009) for Ambulance Trust leaders lists collaborative and networking capabilities under two broad areas. Aspiring leaders must possess "understanding the importance of both partnership working and how to achieve it" (p. 44) and they must be apt at "building collaborative relationships across the wider health community to establish practical ways of delivering organisational objectives" (p. 45). Likewise, College of Policing (2017) calls aspiring leaders to facilitate collaborating with other forces or cross-sector partners because, "working with others ... can improve insights into future planning needs" and "Wider partnership working will require skills of influence and persuasion as well as skills in building and maintaining relationships" (p. 10). However, beyond suggesting that leaders must possess collaborative or networking skills to succeed in their leadership roles, there is lack of understanding on what specific skills are actually required to facilitate collaborations. However, it is assumed that embracing shared leadership approach could contribute in fostering collaborations (Brain and Owens 2015; Reiner and O'Connor 2015; Davis and Bailey 2018). Considering deep embeddedness of command and control culture within emergency services, particularly in the police forces, and the need for shared or distributed leadership, Herrington and Colvin (2016) exhort police organisations to become 'ambidextrous'. Organisations are ambidextrous when they develop the "ability to both 'run the business' (through exploiting existing ways of doing things to be increasingly effective and efficient) and 'change the business' (through exploring new possibilities and innovations)" (Herrington and Colvin 2016; citing Nieto-Rodrigues 2014). Put simply, changing organisational culture would underpin the drive to attain ambidexterity. However the critical question that needs addressing is, who should have the strategic and operational responsibility to drive the collaborative and integration process.

CONCLUSION

Contemporary literature on strategic partnerships highlights the significance of the concept of collaborative capabilities. An organisation's collaborative capability refers to its ability to form and manage strategic partnerships; hence, it is argued that organisations which possess such capabilities derive more value from relationships, thus enhancing the overall organisational performance (Anand and Khanna 2000; Kale and Singh 2009; Kohtamaki et al. 2018). Coordinational mechanism, communicational mechanism and interpersonal bonding are considered as the three critical components underpinning collaborative capabilities. Although organisations develop their collaborative capabilities over time by internalising learning from their past experiences, nonetheless, they also require deliberate and purposeful actions of the organisational leaders (Zollo et al. 2002), because development of partnering capabilities necessarily requires creation, extension and modification of existing organisational structures and operating routines. In specific context to collaborations in the emergency services, the capacity to bond and develop interpersonal relationship will be central in creating a culture that would foster collaborative capabilities. In this respect, the PCCs, CCs and other senior leaders will play a critical role in breaking the boundaries of professional, cultural and identity separation between the three blue light services (Chapman 2015). Over the last decade the three services have made attempts to embrace the concept of shared leadership, even though there is more evidence of continued acceptance of traditional command and control approaches by the organisational members (see, for instance, Davis and Bailey 2018). Creating a culture of collaboration within the single employer organisation will need alternative approaches to leadership, including some aspects of shared and collective as well as integrative leadership.

In his critique of power relationship between leaders and organisational members, Collinson (2011) asserts that "(power relations) are interdependent as well as asymmetrical, typically ambiguous frequently shifting, potentially contradictory and often contested" (p. 185). The notion of power relations, in essence, highlights the centrality of the concept of authority relationship and legitimacy of power structures (see Ladkin 2015) in organisations and organisational forms. The Knight Review (2013) concluded by making a case for wider collaborations between the blue light organisations and called for PCCs to replace the Fire and Rescue Authorities (FRAs). The Policing and Crime Act 2017 enables the PCCs to take on the responsibilities for the fire and rescue services in their areas and places a statutory obligation on the blue light services to collaborate. The resulting image of the single employer model is an organisation that provides operational police and fire functions, headed by a chief constable. In that respect, the PCCs and chief constables (CCs) both have a role in overseeing and facilitating integration of the police and fire and rescue services. Until the creation of PCCs in November 2012, the local democratic accountability of the police was in the hands of police authorities and since then the PCCs have the assumed that role. Hence, the relationship between the PCCs and CCs underpins the effectiveness and efficiency of the single employer model. The Policing Protocol, 2011, sets out how the functions of the PCCs and CCs will be exercised in relations to each other. It envisages that "a constructive working relationship is more likely to be achieved where communication and clarity of understanding are at their highest. Mutual understanding of, and respect for, each party's statutory functions will serve to enhance policing for local communities" (p. 2). However, the critical question to ask here is, what does 'where communication and clarity of understanding are at their highest' stand for? A careful engagement with the Policing Protocol highlights the use of ambiguous language and terminologies that will inevitably create tensions between the two offices. For instance, it states that the CC will decide on "the configuration and organisation of policing resources" (p. 6) but PCC will "decide how the money is allocated" and "decide the budget, allocating assets and funds to the CC" (p. 3). The protocol envisages CC to "remain operationally independent in the service of the communities they serve" (p. 4) and "supporting the PCC in the delivery of the strategy and objectives set out in the plan" (p. 4). Clearly, the PCCs occupy a higher hierarchical position and yet they cannot interact with the subordinates of the CCs, without eroding, if not directly, the authority of the CC and that, one could argue, would hinder the capacity of the PCCs to integrate the police and fire and rescue services within the single employer organisation. Put simply, although demarcation of the hierarchical structure is clear, there is lingering ambiguity in their role as 'leaders' of the single employer organisation. In some instances they could take over the role of the Fire and Rescue Authority (FRA) and hire a CC and a Chief Fire Officer (CFO), and yet the Policing and Crime Act 2017 does not categorically highlight who has the responsibility to take the ownership of creating an integrated, collaborative organisation where the three blue light organisations could co-exist and co-function seamlessly. The inference one could make here is that since the operational responsibilities lie with the CCs and the CFOs towards their respective forces, they might have the responsibilities to create structural and process requirements to function collaboratively. Either way, the single employer model necessitates that the senior leadership, comprising the PCCs and the CCs, CFOs and the leaders of the ambulance services, takes the onus in creating collaborative capabilities in their respective organisations.

References

Alyn, K. (2010, September). Transformational leadership in the fire services: Identifying the needs, motives and values of leaders and flowers. *Fire House*. Retrieved from http://www.firehouse.com/article.10467048/transformational-leadership-in-the-fire-service on 10/02/2019.

Anand, B. N., & Khanna, T. (2000). Do firms learn to create value? The case of alliances. *Strategic Management Journal, 21*, 295–315.

Arikan, A. M., & McGahan, A. M. (2010). The development of capabilities in new firms. *Strategic Management Journal, 31*, 1–18.

Arino, A., de la Torre, J., & Ring, P. S. (2005). Relational quality and inter-personal trust in strategic alliances. *European Management Review, 2*, 15–27.

Bass, B. M. (1985). *Leadership and performance.* New York: Free Press.

Bass, B. M. (1990a). *Bass and Stogdill's handbook of leadership* (3rd ed.). New York: Free Press.

Bass, B. M. (1990b). From transactional to transformational leadership: Learning to share the vision. *Organizational Dynamics, 18*(3), 19–31.

Bass, B. M. (1998). *Transformational leadership: Industrial, military, and educational impact.* Mahwah, NJ: Erlbaum.

Bass, B. M., & Bass, R. (2009). The Bass handbook of leadership: Theory, research, and managerial applications. (4th ed.). New York: Free Press.

Bass, B. M., Avolio, B. J., & Atwater, L. (1996). The transformational and transactional leadership of men and women. *Applied Psychology, 45,* 5–34.

Becker, M. C., & Knudsen, T. (2005). The role of routines in reducing pervasive uncertainty. *Journal of Business Research, 58,* 746–757.

Benne, K. D., & Sheats, P. (1948). Functional roles of group members. *Journal of Social Issues, 4,* 41–49.

Bolden, R., Petrov, G., & Gosling, J. (2008). Tensions in higher education leadership: Towards a multi-level model of leadership practice. *Higher Education Quarterly, 62,* 358–376.

Brain, T., & Owens, L. (2015). Leading in Austerity. In J. Fleming (Ed.), *Police leadership: Rising to the top.* Oxford: Oxford University Press.

Bullock, K., Erol, R., & Tilley, N. (2006). *Problem-oriented policing and partnerships.* Cullompton: Willan Publishing.

Burns, J. M. (1978). *Leadership.* New York: Harper & Row.

Butterfield, L. D., Borgen, W. A., Amundson, N. E., & Maglio, A. T. (2005). Fifty years of the critical incident technique: 1954–2004 and beyond. *Qualitative Research, 5,* 475–497.

Cartwright, S., & Cooper, C. L. (1990). The impact of mergers and acquisitions on people at work: Existing research and issues. *British Journal of Management, 1*(2), 65–76.

Chapman, S. (2015). Crossing cultural boundaries: Reconsidering the cultural characteristics of police officers and ambulance staff. *International Journal of Emergency Services, 4*(2), 158–176.

Chapman, C., Getha-Taylor, H., Holmes, M. H., Jacobson, W. S., Morse, R. S., & Sowa, J. E. (2016). How is public service leadership is studied: An examination of a quarter century of scholarship. *Public Administration, 94,* 111–128.

College of Policing. (2015). *College of policing analysis: Estimating demand on the police service.* Coventry: College of Policing. Retrieved November 10, 2018, from https://www.college.police.uk/News/College-news/Documents/Demand%20 Report%2023_1_15_noBleed.pdf.

College of Policing. (2017). *Guiding principles for organisational leadership.* College of Policing: Ryton-on-Dunsmore.

Collinson, D. (2011). Critical leadership studies. In A. Bryman, D. Collinson, K. Grint, B. Jackson, & M. Uhl-Bien (Eds.), *The Sage handbook of leadership studies* (pp. 181–194). London: Sage.

Cox, K., Jolly, S., Van Der Staaij, S., & Van Stolk, C. (2018). *Understanding the drivers of organisational capacity*. RAND Corporation and Saatchi Institute. Retrieved February 20, 2019, from https://www.rand.org/pubs/research_reports/RR2189.html.

Crosby, B. C., & Bryson, J. M. (2005). A leadership framework for cross-sector collaboration. *Public Management Review, 7*(2), 177–201.

Crosby, B. C., & Bryson, J. M. (2010). Integrative leadership and the creation and maintenance of cross-sector collaborations. *The Leadership Quarterly, 21*(2), 211–230.

Crosby, B. C., & Bryson, J. M. (2018). Why leadership of public leadership research matters: And what to do about it. *Public Management Review, 20*(9), 1265–1286.

Das, T. K., & Teng, B.-S. (2000). Instabilities of strategic alliances: An internal tensions perspective. *Organization Science, 11*(1), 77–101.

Davis, A., & Thomas, R. (2003). Talking cop: Discourses of change and policing identities. *Public Administration, 81*(4), 681–699.

Davis, C., & Bailey, D. (2018). Police leadership: The challenges for developing contemporary practice. *International Journal of Emergency Services, 7*(1), 13–23.

de Rond, M., & Bouchikhi, H. (2004). On the dialectics of strategic alliances. *Organization Science, 15*(1), 56–69.

Department of Health. (2011). *Clinical leadership competency framework*. Coventry: NHS Leadership Academy. NHS Institute for Innovation and Improvement.

Derue, D. S., Nahrgang, J. D., Wellman, N., & Humphrey, S. E. (2011). Trait and behavioural theories of leadership: An integration and meta-analytic test of their relative validity. *Personnel Psychology, 64*, 7–52.

Devos, G., Tuytens, M., & Hulpia, H. (2014). Teachers' organizational commitment: Examining the mediating effects of distributed leadership. *American Journal of Education, 120*(2), 205–231.

Downton, J. V. (1973). *Rebel leadership: Commitment and Charisma in the revolutionary process*. New York: Free Press.

Doz, Y. (1996). The evolution of cooperation in strategic alliances: Initial conditions or learning processes. *Strategic Management Journal, 17*(Summer issue), 55–83.

Drodge, E. N., & Murphy, S. A. (2002). Interrogating emotions in police leadership. *Human Resource Development Review, 1*(4), 420–438.

Dyer, J., & Singh, H. (1998). The relational view: Cooperative strategy and sources of inter-organisational competitive advantage. *Academy of Management Review, 23*, 660–679.

Feldman, M. S., & Pentland, B. T. (2003). Reconceptualising organizational routines as a source of flexibility and change. *Administrative Science Quarterly, 48*, 94–118.

Fiedler, F. E. (1997). Situational control and a dynamic theory of leadership. In K. Grint (Ed.), *Leadership: Classical, contemporary and critical approaches*. Oxford: Oxford University Press.

Getha-Taylor, H., Holmes, M. H., Jacobson, W. S., Morse, R. S., & Sowa, J. W. (2011). Focusing the public leadership lens: Research propositions and questions in the Minnowbrook tradition. *Journal of Public Administration Research and Theory, 21*(1), 83–97.

Gibb, C. A. (1954). Leadership. In G. Lindzey (Ed.), *Handbook of social psychology* (Vol. 2, pp. 877–917). Reading, MA: Addison-Wesley.

Gronn, P. (2008). The future of distributed leadership. *Journal of Educational Administration, 46*(2), 141–158.

Gulati, R. (1998). Alliances and networks. *Strategic Management Journal, 19*(4), 293–317.

Gulati, R., Wohlgezogen, F., & Zhelyazkov, P. (2012). The two facets of collaboration: Cooperation and coordination in strategic alliances. *Academy of Management Annals, 6*, 531–583.

Heikka, J. (2014). *Distributed pedagogical leadership in early childhood education.* Tampere: Tampere University Press.

Heimeriks, K. H., & Duysters, G. (2007). Alliance capability as a mediator between experience and alliance performance: An empirical investigation into alliance capability development process. *Journal of Management Studies, 44*(1), 25–49.

Herrington, V., & Colvin, A. (2016). Police leadership for complex times. *Policing: A Journal of Policy and Practice, 10*(1), 7–16.

Hersey, P., Blanchard, K. H., & Johnson, D. E. (2008). *Management of organizational behavior* (9th ed.). Upper Saddle River, NJ: Pearson Prentice Hall.

HM Government. (2013). *Emergency response and recovery non-statutory guidance to complement emergency preparedness* (online). Retrieved May 04, 2017, from https://www.gov.uk/guidance/emergency-response-and-recovery.

Holt, D., Palmer, S., Gosper, M., Sankey, M., & Allan, G. (2014). Framing and enhancing distributed leadership in the quality management of online learning environments in higher education. *Distance Education, 35*(3), 382–399.

Huxham, C., & Vangen, S. (2005). *Managing to collaborate: The theory and practice of collaborative advantage.* Abingdon, UK: Routledge.

Ireland, R. D., Hitt, M. A., & Vaidyanath, D. (2002). Alliance management as a source of competitive advantage. *Journal of Management, 28*(3), 413–446.

JESIP. (2016, July). *Joint Doctrine – The interoperability framework* (2nd ed.). London: Joint Emergency Services Interoperability Principles (JESIP).

Kale, P., & Singh, H. (2007). Building firm capabilities through learning: The role of the alliance learning process in alliance capability and firm-level alliance success. *Strategic Management Journal, 28*(10), 981–1000.

Kale, P., & Singh, H. (2009). Managing strategic alliances: What do we know now, and where do we go from here? *Academy of Management Perspectives* (August Issue), 45–62.

Kale, P., Dyer, J. H., & Singh, H. (2002). Alliance capability, stock market response, and long term alliance success: The role of the alliance function. *Strategic Management Journal, 23*, 747–767.

Kelloway, K. E., & Barling, J. (2000). What we have learned about developing transformational leaders. *Leadership and Organization Development Journal, 21*(7), 355–362.

Knight, K. (2013). *Facing the future: Findings from the review of efficiencies and operations in fire and rescue authorities in England*. London: HSMO.

Kohtamaki, M., Rabetino, R., & Moller, K. (2018). Alliance capabilities: A review and research agenda. *Industrial Marketing Management, 68*, 188–201.

Ladkin, D. (2015). Leadership, management and headship: Power, emotions and authority in organisations. In B. Carroll, J. Ford, & S. Taylor (Eds.), *Leadership: Contemporary critical perspectives* (pp. 3–25). London: Sage.

Leithwood, K., Mascall, B., & Strauss, T. (2009). *Distributed leadership according to the evidence*. Abingdon: Routledge.

Lewin, K. (1947). Frontiers in group dynamics: Concept, method and reality in social science; equilibrium and social change. *Human Relations, 1*(1), 5–41.

Liu, Y., Sarala, R. M., Cooper, C., & Xing, Y. (2017). Human side of collaborative partnerships: A micro-foundational perspective. *Group & Organization Management, 42*(2), 151–162.

Loftus, B. (2010). Police occupational culture: Classic themes, altered times. *Policing and Society, 20*(1), 1–20.

Lord Kerslake. (2017). The Kerslake Report: An independent review into the preparedness for, and emergency response to, the Manchester Arena attack on 22nd May 2017. Retrieved February 20, 2019 from: https://www.jesip.org.uk/uploads/media/Documents%20Products/Kerslake_Report_Manchester_Are.pdf

Luo, Y. (2006). Opportunism in cooperative alliances: Conditions and solutions. In O. Shenkar & J. J. Reuer (Eds.), *Handbook of strategic alliances* (pp. 55–80). Thousand Oaks, CA: Sage.

Maslow, A. H. (1954). *Motivation and personality*. New York: Harper & Row.

Mohr, J., & Spekman, R. (1994). Characteristics of partnership success: Partnership attributes, communication behavior, and conflict resolution techniques. *Strategic Management Journal, 15*, 135–152.

Morse, R. S. (2008). Developing public leaders in an age of collaborative governance. In R. S. Morse & T. F. Buss (Eds.), *Innovations in public leadership development*. London: M. E. Sharpe.

Morse, R. S. (2010). Integrative public leadership: Catalyzing collaboration to create public value. *The Leadership Quarterly, 21*(2), 231–245.

NHS Ambulance Chief Executive Group. (2009). *Future leaders study: The leadership capabilities and capacities of ambulance trusts in England*.

Northouse, P. G. (1997). *Leadership theory and practice*. Thousand Oaks, CA: Sage.

O'Neill, M., & McCarthy, D. J. (2014). (Re) negotiating police culture through partnership working: Trust, compromise and the 'new' pragmatism. *Criminology & Criminal Justice, 14*(2), 143–159.

Page, S. (2010). Integrative leadership for collaborative governance: Civic engagement in Seattle. *The Leadership Quarterly, 21*(2), 246–263.

Pillai, R., & Williams, E. A. (2004). Transformational leadership, self-efficacy, group cohesiveness, commitment, and performance. *Journal of Organizational Change Management, 17*(2), 144–159.

Reiner, R. (2010). *The politics of the police* (4th ed.). Oxford: Oxford University Press.

Reiner, R., & O'Connor, D. (2015). Politics and policing: The terrible twins. In J. Fleming (Ed.), *Police leadership: Rising to the top* (pp. 42–70). Oxford, UK: Oxford University Press.

Ring, P. S., & Van de Ven, A. H. (1994). Developmental processes of cooperative inter-organizational relationships. *Academy of Management Review, 19*(1), 90–118.

Schreiner, M., Kale, P., & Corsten, D. (2009). What really is alliance management capability and how does it impact alliance outcomes and success? *Strategic Management Journal, 30*, 1395–1419.

Schreyögg, G., & Kliesch-Eberl, M. (2007). How dynamic can organizational capabilities be? Towards a dual-process model of capability dynamization. *Strategic Management Journal, 28*, 913–933.

Shivers-Blackwell, S. (2006). The influence of perceptions of organizational structure & culture on leadership role requirements: The moderating impact of locus of control & self-monitoring. *Journal of Leadership & Organizational Studies, 12*(4), 27–49.

Simonin, B. L. (1997). The importance of collaborative know-how: An empirical test of the learning organization. *Academy of Management Journal, 40*(5), 1150–1174.

Spekman, R. E., Isabella, L. A., & MacAvoy, T. C. (1996). Creating strategic alliances which endure. *Long Range Planning, 29*(3), 346–357.

Sullivan, H., Williams, P., & Jeffares, S. (2012). Leadership for collaboration. *Public Management Review, 14*(1), 41–66.

Taylor, J., & Armitage, E. (2012). Leadership within the ambulance service: Rhetoric or reality? *Journal of Paramedic Practice, 4*(10), 564–568.

The Policing and Crime Act. (2017). *UK Parliament.*

Van Maurik, J. (2001). *Writers on leadership.* Harmondsworth: Penguin.

Van Wart, M. (2003). Public-sector leadership theory: An assessment. *Public Administration Review, 63*, 214–228.

Van Wart, M. (2013). Lessons from leadership theory and the contemporary challenges of leaders. *Public Administration Review, 73*, 553–565.

Vangen, S. (2017). Developing practice-oriented theory on collaboration: A paradox lens. *Public Administration Review, 77*, 263–272.

West, M., Lyubovnikova, J., Eckert, R., & Denis, J. L. (2014). Collective leadership for culture of high quality healthcare. *Journal of Organisational Effectiveness: People and Performance, 1*(3), 240–260.

Wilson, J. M., & Grammich, C. A. (2017). Consolidation of police and fire services in the United States. *International Criminal Justice Review, 27*, 203–221.

Wright, K. (2000). Competency development in public health leadership. *American Journal of Public Health, 90*(8), 1202–1207.

Yu-Lee, R. T. (2002). *Essentials of capacity management.* New York: Wiley.

Yulk, G. A. (2010). *Leadership in organizations* (7th ed.). Upper Saddle River, NJ: Prentice Hall.

Zollo, M., Reuer, J. J., & Singh, H. (2002). Inter-organizational routines and performance in strategic alliances. *Organization Science, 13*, 701–713.

CHAPTER 4

Trust in the Context of Emergency Service Collaborations

Abstract This chapter concerns with exploring the significance and various dimensions of trust in the operation and functioning of inter-organisational partnerships. Trust is one of the most studied themes in the domain of inter-organisational collaborations and it is considered to contribute to enhancing the quality and level of cooperation between the partners, thus lowering the governance and transactional cost, reducing conflicts, increasing partner predictability and facilitating learning and knowledge exchange by paving way for development of informal and interpersonal relationships. Therefore, in many respects, trust is an element and outcome of success (or failure) in inter-organisational relationships. However, integration of emergency services is not merely a collaborative process but also a change process; therefore, exploring inter-organisational trust would also necessarily involve accounting for trust relationship within the partner organisations and its implication on the overall strategic partnership. In this context, the concept of psychological contract assumes relevance.

Keywords Blue light collaborations • Tensions and instability • Inter-organisational trust • Multilevel and multidimensional trust • Psychological contract

© The Author(s) 2020
P. Wankhade, S. Patnaik, *Collaboration and Governance in the Emergency Services,*
https://doi.org/10.1007/978-3-030-21329-9_4

INTRODUCTION

Extant literature on inter-organisational relationships sensitises us to the view that these organisational arrangements are not naturally occurring phenomena (Spekman et al. 1998). In fact, left on their own, managers would rather prefer to minimise their level of dependence on others and not share their decision-making privilege with colleagues from partner organisations. Simply put, organisations are essentially organised to undertake specific tasks and activities within their own respective organisational boundaries and therefore forming strategic partnerships with other organisations add to existing organisational as well as managerial complexities (Gulati and Singh 1998; Bell et al. 2006). Addition of organisational and managerial complexities creates conditions for tensions between the partners. Hence, inter-organisational partnerships are also considered as a contested terrain that involves tensions and instabilities, which eventually contribute to their failure and termination. It is in this backdrop that trust has emerged as one of the central themes in research on success and failure of inter-organisational relationships (Zaheer and Harris 2006; Schilke and Cook 2013). Put simply, trust is also conceptualised as a mechanism for dealing with tensions and uncertainties that underpin inter-organisational relationships. However, the research on trust in inter-organisational partnership is fragmented because prior studies have examined this phenomenon at varying levels, either at individual or at organisational levels. However, increasingly there have been calls to conceptualise trust as a multidimensional, multilevel dynamic construct that develops over a period of time.

In this chapter, we aim to provide a detailed and yet nuanced understanding of trust in the context of inter-organisational relationships. First, we delineate the tensions and forces that contribute to creating instabilities in strategic partnerships. Then we provide a broad overview of how trust is conceptualised in inter-organisational relationship literature and identify its various dimensions. Considering that public sector organisations in general and the emergency services in particular have borne the blunt of austerity-centric reforms, it is imperative that changing intra-organisational dynamics is also accounted for, and in this respect we explore the concept of psychological contract that pertains to the nature of relationship between the employees and the employers. We conclude the chapter with some reflection on trust in the context of the collaborations amongst the blue light organisations.

TENSIONS AND INSTABILITIES IN INTER-ORGANISATIONAL COLLABORATIONS

Das and Teng (2000) were amongst the first who suggested that inter-organisational partnerships should be conceptualised as organisational arrangements where competing forces vie to co-exist. In this context, they identified three pairs of competing forces—cooperation versus competition, rigidity versus flexibility and short-term versus long-term orientations—as the source of tensions and conflicts in partnerships.

Cooperation Versus Competition Clearly, cooperation and competition are opposing forces that characterise various partnerships. Whereas cooperation is underpinned by goodwill, collective interests and common benefits, competition is reflected in opportunistic behaviour, withholding knowledge and information and seeking private benefits. Even when pursuing collaborations, partner organisations continue to compete. Considering collaborations amongst emergency services involve organisations with long cherished history and established work/organisational practices, and decision to collaborate is underpinned by institutional, resource and efficiency considerations, it is but obvious that tensions would result from who gets more access to resources and whose practices should be adopted as the partnership operationalises. In the context of collaborations between the blue light organisations, even though the cooperation–competition paradox does not seem to be relevant, in many respect it actually would underpin the relationship. The centrality of achieving efficiency underpins the drive for better (and more) collaborations between the police and fire services in particular (see Policing and Crime Act 2017). This creates conditions for competition amongst the two services to vie for funding resources, at least in the short term until governance structure (either under a single employer model or under a PCC-FRA model) is institutionalised.

Rigidity Versus Flexibility Das and Teng (2000) present rigidity as a structural feature that highlights how (a) elements within organisations are linked to one another and (b) organisational elements within one partner link to organisational elements of the partnering organisations. However, insights from behavioural studies (e.g. Rosman et al. 1994) suggest that rigidity and flexibility are not merely structural constructs but also behavioural constructs, which help in explaining organisational decisions in different contexts, including strategic alliances. Structural rigidity and behaviour rigidity, in strategic alliances, create hindrances on how the relationships develop over time (Ring and Van de Ven 1994; de Rond and Bouchikhi 2004; Bryson et al. 2015). Structural flexibility and behavioural flexibility, instead, are critical for partnerships to survive and succeed since these arrangements necessarily have to co-evolve with changes in the internal and external dynamics of the partner organisations (Koza and Lewin 1998). Although all the three blue light organisations come under the emergency service category, they nevertheless perform different roles at different points in time. However, each of the organisations has its own respective routines which, even in the emergency situations, they rigidly hold on to and that have significant implications on the delivery of the service. In this respect, The Kerslake Report (2017) highlights significant gaps in the collaborative functioning of the three services and in particular it refers to how the operational protocols that each organisation follows created rigidities at a time that demanded flexibility.

Long-Term Versus Short-Term Orientation One of the fundamental challenges that strategic partnerships face relates to the ambiguity in terms of their duration of existence. In other words, it is not always known beforehand either how long would a decision to enter into an alliance last or how long decisions pertaining to structural configurations of the partnership itself would last. Lack of long-term orientation to different decisions relating to collaborations creates tensions and instability. This is particularly critical because, in contrast to partnerships in the public sector context, collaborations in public sector, more so in emergency services, are, in essence, change initiatives that aim to bring about changes in how the police forces, ambulance services and fire and rescue service work on their own as well as in partnership with each other (Policing and Crime Act 2017). Although in the long term it is envisaged that of the three, two services, the police service and the fire and rescue service, would come under the single employer model of governance, it still remains a long process. As a result, in the short term, the governing structure for day-to-day functioning needs significant clarity for any collaboration to effectively deliver efficiency. It is in this respect, that is, lack of structural clarity, that trust between the services assumes significance.

Trust as an Antidote

Considering inter-organisational partnerships are rife with tensions and instability, resulting from co-existence of contradictory forces, trust is often conceptualised as the antidote to facilitate smoother functioning of the relationships. Trust, in general, is one of the most studied concepts of social sciences due to the role it plays in the construction and functioning of social life; yet there is no one universally accepted definition of trust. Rousseau et al. (1998), in their study on cross-disciplinary collection of scholarly work on trust, identified a widely held definition of trust, which goes as follows: "Trust is a psychological state comprising the intention to accept vulnerability based on positive expectation of the intensions or behaviour of another" (1998, pp. 395). This definition identifies 'risk' and 'interdependence' as the two necessary preconditions that must exist for trust to arise. Das and Teng (2000) differentiate between objective and subjective risks. Whereas objective risk is a measurable consequence or outcome of alternatives and their probabilities, subjective risk (also known as perceived risk) is the probability of loss, as perceived by the social actors. Trust plays a central role in reducing perceived risk. Interdependence, highlights that the interest of one entity (individual or organisation) cannot be achieved without reliance upon another entity. It also emphasises existence of reciprocal relationships as central to formation and maintenance of trust. Thus, interdependence is one of the most distinctive features of inter-organisational partnerships. In specific context to blue light organisations, the principles of interoperability is underpinned by the notion of interdependence. In making the case for more collaborative functioning amongst the blue light organisations, the Knight Review (2013) alludes to potential areas of interdependence that exist for delivering the services.

This widely held definition also highlights a critical aspect of trust, that is, in essence, it is an interpersonal-level construct. Interestingly although trust exists between individuals, most research on inter-organisational trust considers organisations as synonymous to a single actor. Simply put, existence of trust between partner organisations is a reflection of trust between individuals belonging to the partner organisations (Vanneste 2016). Therefore, the broad question that arises is—should the trust relationship between individuals at senior management level be considered at par with the trust relationship between individuals at, perhaps, lower hierarchical levels? This question is critical because smooth functioning of inter-organisational relationships necessarily entails involvement of individuals in different hierarchical levels, even though, interestingly, within trust research, emphasis is given to perception of 'decision makers' (Rousseau et al. 1998; Das and Teng 2000). This question is also relevant in the case of strategic partnership involving the blue light organisations, where the notion of interoperability highlights the centrality of joint operations, which in practice involve participation of members of the operational teams, who are often considered occupying a lower hierarchical space as compared to those involved in formulating strategies (see Faems et al. 2008; also see Charman 2017). Notwithstanding these differences, there is broad consensus that trust facilitates cooperation between the partners as well as contributes towards enhancing the quality of their relationship, lowering governance cost, reducing conflicts, increasing partner predictability, improving flexibility amongst the partners and facilitating learning and knowledge exchange.

As we noted earlier in the chapter, risk is a condition for existence of trust. In specific context to inter-organisational relationships, the contradictory co-existing forces create conditions for development of risk perception. Hence, the assertion that understanding the risks involved in inter-organisational relationships is central to understanding how trust manifests in and facilitates functioning of strategic partnerships, such as the proposed collaboration between the blue light organisations. Inkpen and Currall (2004) define risk in strategic partnerships as, "(is the) potential that the trusting party will experience negative outcomes, if the other party proves untrustworthy, and *therefore perception of* risk creates the opportunity for trust (*or distrust*)" (pp. 588; *italics added*). Risks, in inter-organisational relationships, are primarily of two types, namely relational risk and performance risk.

Relational Risk In the context of inter-organisational relationships, relational risk pertains to the risk that a partner would not cooperate satisfactorily and this is particularly associated with opportunistic behaviour of partner organisations. The key argument that underpins the opportunistic behaviour of partners is that irrespective of the decision of the organisations to collaborate, they still retain and nurture their respective self/individual interests. Therefore, there is possibility that partners' actions could be driven more towards attaining private benefit as compared to common benefit. Therefore, it is critical that 'common purpose' should act as a glue to mitigate opportunistic behaviour of partners.

Existing research from the domain of public management illuminates the centrality of a higher purpose that drives commitment of public sector employees amidst cuts and uncertainty (Vangen and Huxham 2003; Conway et al. 2014).

Performance Risk In inter-organisational relationships, cooperative behaviour of the partners does not necessarily lead to desirable performance. Therefore, performance risk, which pertains to the possibility that the partners might not fulfil their overarching objectives in spite of their best intentions and efforts, is a distinctive feature of collaborations. Various factors—including structural and personnel changes within the partner organisations, distinctly different work practices and organisational cultures, lack of expertise or competence and even changes in the institutional/government policies—contribute to partners' inability to achieve desirable objectives. In this context, it is critical to highlight that risks—relational or performance—are essentially perceptions and the perceptions are shaped by social discourse (see Fairclough 1993, 2003; Cerna, 2013) of critical events. In her interesting work on the relationship between police and ambulance service staffs, Charman (2015) refers to how they positively viewed each other. She notes, "Ambulance staff were described by police officers as being 'in tune' (R1) with police, 'good as gold' (B3), 'like-minded people' (B4) and 'it sounds corny [...] they're great' (E4). Ambulance staff described a relationship that was 'incredibly collaborative' (P3), had a 'natural affinity' (P19), 'a huge amount of reciprocal respect' (P3) and was 'friendly and fun' (P20)" (Charman 2015, p. 163). Having good opinion about each other reflects their perception of each other's expertise and competence.

MULTIDIMENSIONAL NATURE OF TRUST

The presence of interdependence and risks in inter-organisational relationships highlights the multidimensional nature of trust. Trust, most scholars agree, is underpinned by personal characteristics of individuals involved in the formation and operationalisation of partnerships, institutional context and situational factors (Zaheer and Harris 2006; Vanneste 2016). Das and Teng (2001) and Nooteboom (1996) proposed that trust is of two types, namely (a) goodwill trust and (b) competence trust. Both these types of trust influence the capacity of an organisation to become 'partner of choice'.

Goodwill Trust Considering opportunistic behaviour is a distinctive feature of inter-organisational relationships, there is always a question mark on the intentions of the partners to undertake and jointly perform agreed tasks. In other words, the possibility partners would indulge in activities that would benefit them at the expense of the partnership is present in all such relationships. Therefore, goodwill trust pertains to a partner's reputation regarding good faith, intentions and integrity. Having such reputation, in essence, reduces relational risk. Many scholars, who take a process view of trust development, argue that goodwill trust reflects prior behaviour of a partner in other collaborations.

Competence Trust In contrast to goodwill trust, competence trust pertains to the ability of the partners to perform as per the agreed terms and conditions. Competence is based on organisational resources, such as human capital, technology and expertise, and capabilities that an organisation possesses and thus provides the basis for an alliance to achieve its objectives. Competences of partners are also reflected in their prior achievements and, therefore, it assumed that high competence of partners in an inter-organisational relationship lowers the performance risk.

In specific context to goodwill and competence trust in case of collaborations between the blue light organisations, the Emergency Service Collaboration Working Group's national overview (2016) specifically highlights some of the existing collaborations between the ambulance services and the fire and rescue services in some areas and between the police and ambulance services and police and fire services in other areas. In most instances, the overview (2016) suggests improvements in effectiveness and efficiency savings. The illustrations provided in the overview, in essence, capture the efforts to develop both goodwill and competence trust.

However, trust is not a static construct; rather, it is dynamic in the sense that trust evolves as relationships evolve. Zucker (1987) was amongst the first to highlight how trust develops in various exchange relationships and suggested three factors that underpin development of trust in social contexts. *Process-based trust* develops from long-term, largely stable relationships wherein each partner holds the view that the other partner would continue to behave as it had done in the past. Simply put, repeated interaction creates conditions for development of trust (Zaheer and Harris 2006; Gulati and Sytch 2008). How trust also develops depends on the *characteristic features* of the partners. Common aspects such as common nationality, religion, family or ethnic backgrounds also create conditions for development of trust. Apart from the process-based and characteristic-based trust, Zucker (1987) also emphasised the significance of embeddedness of organisations in institutional environment in creating conditions for development of trust. The *institutional context* pertains to the existence of formal structures within the social system and it provides basic but influential foundation to how partners come together and engage with each other. The common thread that runs through the three factors is that 'familiarity breeds trust' (Gulati 1995). Seen through the lens of blue light collaborations, all the three factors, identified by Zucker (1987), are relevant, specifically the institutional context and the characteristics features. All the three types of organisations, police forces, ambulance services and fire and rescue services, are embedded within the same broad institutional context and they share the same distinctive characteristics, that is, they are involved in providing public services. In other words, these features ensure that the possibility of establishing trust relationship between organisations and organisational members would be relatively feasible as compared to collaborations between the blue light organisations and non-blue light organisations. However, that does not diminish the possibilities of misunderstanding due to lack of communication and knowledge about each other's existing operational routines or protocols. In this respect, the Kerslake Report (2017) on Manchester Arena bombing illuminates the communication and coordination challenges.

Trust building, as Zhang and Huxham (2009) note, is a nuanced and complex process. Elsewhere Vangen and Huxham (2003), building on their longitudinal research on strategic inter-organisational collaborations in public sector organisations (also see Vangen and Huxham 2013) and by adopting process perspective, conceptualise development of inter-organisational trust as a cyclical process that begins with the collaborating partners aiming to achieve some realistic, in fact, initially modest, outcomes. How partners attain success in achieving the outcomes enforces trust relationship, which allows them to set more challenging and ambitious objectives. Put simply, expectation formation and fulfilment of the expectations underpin the development and maintenance of trust. Each time the partners decide to work together to achieve some objectives, each partner is undertaking some degree of risk; thus, the cyclical conceptualisation of trust development is built on the notion of goodwill and performance risk. Two elements are central to the trust development process, namely repeated evaluation of each other's expertise and behaviour and social interaction between the key actors. The process approach of conceptualising trust essentially builds using the thematic tools posited by Ring and Van de Ven (1994), which suggested that partners repeatedly evaluate each other's intentions and past behaviour as inter-organisational collaborations develop over time. The Emergency Service Collaboration Working Group's National Overview (2016) provides a list of approximately more than 40 ongoing collaborations between various blue light organisations, nationwide. Although the overview delineates specific benefits accrued by the respective partnerships, there is less insight on how they have achieved closer functioning and, in that respect, longitudinal studies on some of these existing and ongoing collaborations could shed light on how the trust relationship was formed, developed and maintained.

The notion that trust develops over time, underpinned by cultural and institutional familiarity and the repeated cycle of expectation formation and fulfilment, brings to forth centrality of (a) key actors or individuals, also known as boundary spanners and (b) non-boundary spanners. The emphasis on individuals, as boundary spanners or otherwise, signify the criticality of interpersonal relationships in the trust-building process. As Vanneste (2016) rightly notes, when it comes to trust, "it is people who trust and not organisations" (p. 7). Simply out, the trust between partner organisations is a result of trust between boundary spanners and others. In this respect, previous studies have highlighted the positive relationship between interpersonal relationship and performance of strategic partnerships. Boundary spanners, also known as alliance managers (Kale and Singh 2009), are individuals whose primary role is to perform day-to-day management of inter-organisational relationships. In essences, their role entails them functioning across organisational boundaries and acting as the connection between the collaborating partners. At the same time, the boundary spanners face considerable challenges in dealing with tensions resulting from lack of clarity and multiple accountabilities, differing cultural systems and governance structures (Williams 2013). These challenges create hindrance in flow of information and knowledge, coordination of activities and facilitation of functioning relationships at different hierarchical levels. Therefore, it is critical to account for the backdrop and context of collaborations to develop a better understanding of how trust is developed and maintained in inter-organisational relationships. In a debate in the House of Commons on the Police and Crime Bill, the then Home Secretary praised the collaborative work of Hampshire Fire and Rescue

Services (HFRS) and Hampshire Constabulary (HC), which was formed under a pilot scheme in 2015–2016, as an illustration of a collaboration that has successfully attained considerable degree of interoperability. In a press release Assistant Chief Officer Stew Adamson (HFRS 2017), who oversaw formation and development of the collaboration, provided some insights on how the collaboration came about and the structural and relational mechanisms the two organisations have incorporated to make the partnership work. For instance, HFRS first invited the HC to use their headquarters to co-share some of the facilities. That apart, they also made changes to their composition of the operational team, in the sense that a police officer was added to the HFRS-run Arson Task Force, organising cross-servicing training to 'upskill firefighters and police officers'. The HC also reciprocated by including one of the firefighters 'going on a beat' with the PC. Apart from these operation-related activities, the HFRS also invited the Police and Crime Commissioners (PCCs) to attend numerous meetings. Thus a combination of activities at a structural and personal level was pursued to acquaint to each other's operating routines and develop personal relationships. This is consistent with the arguments we have made earlier that repeated interactions amongst various key individuals and increased participation in each other's operational activities create familiarity, leading to greater trust relationship (Gulati and Sytch 2008; Silvia 2018). Interestingly, HFRS had prior ongoing collaboration with South Central Ambulance Service for approximately ten years, and the critical learning from that relationship, an illustration of alliance experience, which we highlighted in the previous chapter, contributed to the development and management of HFRS's collaboration with HC. In this respect, we observe that prior experience of HFRS has contributed in shaping the collaboration and reflects its collaborative capabilities (Kohtamäki et al. 2018; Wang and Rajagopalan 2015; Clarke and MacDonald 2019). The significance of boundary spanner as 'learning' agent is developed in the literature on inter-organisational collaborations (Kale and Singh 2009; Niesten and Jolink 2015). Williams (2013) highlights the critical and difficult roles boundary spanners play in collaborative context and calls for the competencies associated with the role be mainstreamed in different leadership and management practices. Although the focus on the boundary spanners and the nature of their roles and activities is still in its infancy, it is nevertheless a welcome development. Therefore, it is imperative that issues pertaining to key individuals, particularly those who are directly associated with collaborations at strategic and operational levels, need better understanding and examination.

PSYCHOLOGICAL CONTRACT AND TRUST BUILDING

From the earlier description on trust and development of trust, we distil two key points. First, trust development is a complex process that develops over time and second it involves repeated interaction between key individuals, who in turn interact with others within their own respective organisations. Put simply, if trust is critical for a collaboration to develop and succeed, it necessarily entails active engagement of members of the organisations. Emerging studies, albeit in a different context, highlight that organisational members reduce their interaction with others from partner organisations if they perceive changes in power dynamics or power asymmetry (Mahadevan. 2011; Zimmermann and Ravishankar 2014). The critical limitation of this understanding of trust and trust building is the conceptualisation that dynamics within partner organisations have negligible bearing on the functioning of strategic partnerships. This is particularly significant because, as emerging studies highlight, inter-organisational relationships, particularly relating to mergers and acquisitions, also create anxieties, stress and emotional distress to organisational members (Hassett et al. 2018; Kusstatscher and Cooper 2005; Cartwright and Cooper 1994). Blurring of organisational boundaries is one of the possible outcomes of the blue light collaboration, which in turn also creates conditions for emotional upheaval of organisational members. These issues are central to the current operating environment of the blue light organisations and some of the underpinning issues relating to people issues are delineated in the subsequent chapters.

We forward the argument that collaborations, in the context of the blue light organisations, are also organisational change initiatives because collaborations are considered as strategic response to changes in the institutional environment of the organisations, and the collaborations also result in changes in the nature of work environment (see, for instance, Coyle-Shapiro and Kessler 2003; Conway et al. 2014; Rayton and Yalabik 2014; Harrington and Lee 2015). Vast bodies of literature on organisational change have sensitised us to the significance of the concept of psychological contract in shaping the success or failure of change initiatives. Although research on psychological contract and its implications have a long history (Conway et al. 2014; Coyle-Shapiro and Parzefall 2008), it was Rousseau's (1989) reconceptualisation of the concept which is most widely used in contemporary studies. Psychological contract, according to Rousseau (1989), is conceptualised as an organisational member's (employee's) beliefs concerning the mutual obligations that exist between him/her and the employer. Broadly speaking, it is the perception of the employee regarding mutual obligation that exists in his/her engagement with the employer and the obligations are sustained through norms of reciprocity. The notion of reciprocity is also central to the concept of trust (Vanneste 2016). Elsewhere studies on intra-firm trust or employee trust, specifically the trust relationship between employees and management, which mirrors insights from psychological contract, also suggest that decline of trust between employee and management contributes to decline in workforce commitment and overall performance (Brown et al. 2011). Organisational change often results in changes in the employment relationship. Extant studies suggest that employees often consider organisational change initiatives that result in downsizing, outsourcing or major restructuring, as violations of psychological contract. In collaborations amongst the blue light organisations, it is expected to result in either downsizing of roles and positions or major restructuring on how (a) each blue light organisation would function on independently; and (b) how they would function in collaboration with other blue light organisations. In either cases, how the organisational members will perceive their role in the new environment will have significant bearing on the approach they would adopt to collaborative with their counterparts. Therefore, it is imperative that people-specific issues, including resilience and well-being of organisational members and issues pertaining to cultural changes and social identifies, need closer attention of the senior managers.

CONCLUSION

Inter-organisational relationships characterised by tensions and uncertainties suffer from a high rate of failure. In this backdrop, trust assumes significance in the management of strategic partnerships as a mechanism to deal with uncertainties. Although there is greater clarity regarding different types of forces that create conditions for tension and instability in collaborations and potential risks the relationships present to each partners, there is still lack of understanding on how trust could be developed and measured. The central issue here is how trust is conceptualised to develop robust explanations on how they facilitate smooth evolution of inter-organisational partnerships. Clearly, trust is a multidimensional construct and trust building between organisations inevitably involves trust between members of the partner organisations. Only when there is trust relationship between organisational members that trust develops between the partnering organisations. Thus, extant studies on inter-organisational relationships have explored trust, predominantly at either organisational level (organisation–organisation) or at individual level (interpersonal trust between key individuals or boundary spanners). However, strategic partnerships develop over time and so does the trust dynamics and therefore there have increasingly been calls to develop multilevel understanding of how collaborating partners build and develop trust and how they use trust to resolve potentially conflicting situations or address structural and relational tensions.

The call for multilevel explanation of trust building is underpinned by the view that inter-organisational partnerships evolve over time, which means the stages through which the partnership develops are temporal and identifiable. There are distinct set of activities and involvement of different set of individuals and therefore different trust relations (Schilke and Cook 2013). Repeated interaction amongst boundary spanners, at different hierarchical levels, creates conditions that foster trust dynamics between organisations. However, as recent studies highlight, collaborations also create instabilities within partner organisations. More so, in blue light organisations, the drive for collaboration has a strong 'organisational change' and 'organisational restructuring' undertone and it is reflected in the policy documents that have sought to posit the argument that collaborations between the blue light organisations would result in greater efficiency and effectiveness. In other words, the collaborations are viewed as a vehicle to restructure how blue light organisations, on their own and collectively, would deliver service in terms of existing and functioning on their own or delivering collaboratively. In this context, the notion of psychological contract—the perceived relationship between employees and employers—assumes significance. Decline in quality of psychological contract, we argue, would have significant implications on trust relationship between organisational members, resulting in poor relational quality. Another dimension of trust that also needs reflection in the context of the blue light collaborations is the relationship between these organisations, particularly the police forces, and the public at large. Public trust pertains to perception of the public at large and to the competences and effectiveness of emergency service organisations, which would feed into how employees within the organisations perceive the collaborations per se.

References

Bell, J., den Ouden, B., & Ziggers, G. W. (2006). Dynamics of cooperation: At the brink of irrelevance. *Journal of Management Studies, 43*(7), 1607–1619.

Brown, S., McHardy, J., McNabb, R., & Taylor, K. (2011). Workplace performance, worker commitment, and loyalty. *Journal of Economics & Management Strategy, 20*, 925–955.

Bryson, J. M., Crosby, B. C., & Stone, M. M. (2015). Designing and implementing cross-sector collaborations: Needed and challenging. *Public Administration Review, 75*, 647–663.

Cartwright, S., & Cooper, C. (1994). The human effects of mergers and acquisitions. *Journal of Organizational Behaviour (1986–1998), 1*, 47–61 (Trends in Organizational Behavior).

Cerna, L. (2013). *The nature of policy change and implementation: A review of different theoretical approaches.* Geneva: The Organisation of Economic Cooperation and Development.

Charman, S. (2015). Crossing cultural boundaries: Reconsidering the cultural characteristics of police officers and ambulance staff. *International Journal of Emergency Services, 4*(2), 158–176.

Charman, S. (2017). *Police socialisation, identity and culture: Becoming blue.* Basingstoke: Palgrave Macmillan.

Clarke, A., & MacDonald, A. (2019). Outcomes to partners in multi-stakeholder cross-sector partnerships: A resource-based view. *Business & Society, 58*(2), 298–332.

Conway, N., Kiefer, T., Hartley, J., & Briner, R. B. (2014). Change and psychological contract breach. *British Journal of Management, 25*, 737–754.

Coyle-Shapiro, J. A.-M., & Kessler, I. (2003). The employment relationship in the U.K. public sector: A psychological contract perspective. *Journal of Public Administration Research and Theory, 13*(2), 213–230.

Coyle-Shapiro, J. A.-M., & Parzefall, M. (2008). Psychological contracts. In C. L. Cooper & J. Barling (Eds.), *The Sage handbook of organizational behavior* (pp. 17–34). London: Sage.

Das, T. K., & Teng, B.-S. (2000). Instabilities of strategic alliances: An internal tensions perspective. *Organization Science, 11*(1), 77–101.

Das, T. K., & Teng, B.-S. (2001). Trust, control, and risk in strategic alliances: An integrated framework. *Organization Studies, 22*(2), 251–283.

de Rond, M., & Bouchikhi, H. (2004). On the dialectics of strategic alliances. *Organization Science, 15*(1), 56–69.

Faems, D., Janssens, M., Madhok, A., & Van Looy, B. (2008). Toward an integrative perspective on alliance governance: Connecting contract design, trust dynamics, and contract application. *Academy of Management Journal, 51*(6), 1053–1078.

Fairclough, N. (1993). Critical discourse analysis and the marketization of public discourse: The universities. *Discourse & Society, 4*(2), 133–168.

Fairclough, N. (2003). *Analysing discourse: Textual analysis for social research.* London: Routledge.

Gulati, R. (1995). Does familiarity breed trust? The implication of repeated ties for contractual choice in alliances. *Academy of Management Journal, 38*(1), 85–112.

Gulati, R., & Singh, H. (1998). The architecture of cooperation: Managing coordination costs and appropriation concerns in strategic alliances. *Administrative Science Quarterly, 43*(4), 781–814.

Gulati, R., & Sytch, M. (2008). Does familiarity breed trust? Revisiting the antecedents of trust. *Managerial Decision and Economics, 29*, 165–190.

Harrington, J. R., & Lee, J. H. (2015). What drives perceived fairness of performance appraisal? Exploring the effects of psychological contract fulfillment on employees' perceived fairness of performance appraisal in U.S. Federal Agencies. *Public Personnel Management, 44*(2), 214–238.

Hassett, M. E., Reynolds, N.-S., & Sandberg, B. (2018). The emotions of top managers and key persons in cross-border M&As: Evidence from a longitudinal case study. *International Business Review, 27*(4), 737–754.

HFRS. (2017). *Hampshire is beacon of blue light collaboration: Partnership work between fire, police and ambulance benefits the public.* Retrieved February 22, 2019, from https://www.hantsfire.gov.uk/incidents-news-and-events/news-releases/2017/hampshire-is-beacon-of-blue-light-collaboration/.

Inkpen, A. C., & Currall, S. C. (2004). The co-evolution of trust, control, and learning in joint ventures. *Organization Science, 15*(5), 586–599.

Kale, P., & Singh, H. (2009, August). Managing strategic alliances: What do we know now, and where do we go from here? *Academy of Management Perspectives*, 45–62.

Knight Report. (2013). *Facing the future: Findings from the review of efficiencies and operations in fire and rescue authorities in England.*

Kohtamäki, M., Rabetino, R., & Möller, K. (2018). Alliance capabilities: A systematic review and future research directions. *Industrial Marketing Management, 68*, 188–201.

Koza, M. P., & Lewin, A. Y. (1998). The co-evolution of strategic alliances. *Organization Science, 9*(3), 255–264.

Kusstatscher, V., & Cooper, C. (2005). *Managing emotions in mergers and acquisitions.* Cheltenham: Edward Elgar.

Lord Kerslake. (2017). The Kerslake report: An independent review into the preparedness for, and emergency response to, the Manchester Arena attack on 22nd May 2017. Retrieved February 20, 2019, from https://www.jesip.org.uk/uploads/media/Documents%20Products/Kerslake_Report_Manchester_Are.pdf.

Mahadevan, J. (2011). Engineering culture(s) across sites: Implications for cross-cultural management of emic meanings. In H. Primecz, L. Romani, & S. Sackmann (Eds.), *Cross-cultural management in practice: Culture and negotiated meanings* (pp. 89–100). Cheltenham: Edward Elgar.

Niesten, E., & Jolink, A. (2015). Alliance management capabilities and performance. *International Journal of Management Reviews, 17,* 69–100.

Nooteboom, B. (1996). Trust, opportunism and governance: A process and control model. *Organization Studies, 17*(6), 985–1010.

Rayton, B. A., & Yalabik, Z. Y. (2014). Work engagement, psychological contract breach and job satisfaction. *The International Journal of Human Resource Management, 25*(17), 2382–2400.

Ring, P. S., & Van de Ven, A. H. (1994). Developmental processes of cooperative inter-organizational relationships. *Academy of Management Review, 19*(1), 90–118.

Rosman, A., Lubatkin, M., & O'Neill, H. (1994). Rigidity in decision behaviors: A within-subject test of information acquisition using strategic and financial informational cues. *The Academy of Management Journal, 37*(4), 1017–1033.

Rousseau, D. M. (1989). Psychological and implied contracts in organizations. *Employee Responsibilities and Rights Journal, 2,* 121–139.

Rousseau, D. M., Sitkin, S. B., Burt, R. S., & Camerer, C. (1998). Not so different after all: A cross-discipline view of trust. *Academy of Management Review, 23,* 393–404.

Schilke, O., & Cook, K. S. (2013). A cross-level process theory of trust development in interorganizational relationships. *Strategic Organization, 11*(3), 281–303.

Silvia, C. (2018). Evaluating collaboration: The solution to one problem often causes another. *Public Administration Review, 78,* 472–478.

Spekman, R. E., Forbes, T. M., Isabella, L. A., & MacAvoy, T. C. (1998). Alliance management: A view from the past and a look into the future. *Journal of Management Studies, 35*(6), 747–772.

The Policing and Crime Act. (2017). *UK Parliament.* Retrieved January 20, 2018, from http://www.legislation.gov.uk/ukpga/2017/3/pdfs/ukpga_20170003_en.pdf.

Vangen, S., & Huxham, C. (2003). Nurturing collaborative relations: Building trust in interorganizational collaboration. *The Journal of Applied Behavioral Science, 39*(1), 5–31.

Vangen, S., & Huxham, C. (2013). Building and using the theory of collaborative advantage. In R. Keast, M. P. Mandell, & R. Agranoff (Eds.), *Network theory in the public sector: Building new theoretical frameworks* (pp. 51–69). New York: Routledge.

Vanneste, B. S. (2016). From interpersonal to inter-organisational trust: The role of indirect reciprocity. *Journal of Trust Research, 6*(1), 7–36.

Wang, Y., & Rajagopalan, N. (2015). Alliance capabilities: Review and research agenda. *Journal of Management, 41*(1), 236–260.

Williams, P. (2013). We are all boundary spanners now? *International Journal of Public Sector Management, 26*(1), 17–32.

Zaheer, A., & Harris, J. (2006). Inter-organizational trust. In O. Shenkar & J. Reuer (Eds.), *Handbook of strategic alliances.* London: Sage.

Zhang, Y., & Huxham, C. (2009). Identity construction and trust building in developing international collaborations. *The Journal of Applied Behavioral Science, 45*(2), 186–211.

Zimmermann, A., & Ravishankar, M. N. (2014). Knowledge transfer in IT offshoring relationships: The roles of social capital, efficacy and outcome expectations. *Information Systems Journal, 24,* 167–202.

Zucker, L. G. (1987). Institutional theories of organization. *Annual Review of Sociology, 13,* 443–464.

CHAPTER 5

Mental Health and Well-Being of the Emergency Services Workforce

Abstract Cases of stress, poor mental health and post-traumatic stress disorder (PTSD) are on the rise in the emergency service workers. Harassment and bullying instances reported in media are also now included within official reports. Research conducted by the charity *MIND* suggests that nine out of ten members of the emergency services have experienced stress, low mood or poor mental health at some point whilst at work. The chapter aims to explore the issues of sickness absence in the three main emergency services, drawing from the current evidence. It analyses the implications of discrimination and exploration of incidences of bullying and harassment while assessing the impact on staff and organisation. It also highlights initiatives for blue light well-being and developing awareness of factors perpetuating staff sickness and implications for organisational productivity and resilience management. It provides avenues for further research.

Keywords Well-being • Mental health • Emergency services • Stress • PTSD • Sickness absence • Resilience

INTRODUCTION AND BACKGROUND

Staff working across the emergency services sector often work in emotionally charged situations involving work which can be characterised as intense (Granter et al. 2019), often in full public view, thanks to the social media.

© The Author(s) 2020   83
P. Wankhade, S. Patnaik, *Collaboration and Governance in the
Emergency Services*,
https://doi.org/10.1007/978-3-030-21329-9_5

Such work is often mediated with increasing demand, reducing budgets and growing levels of stress and mental health issues. According to *MIND*, the mental health charity, more than 85% of emergency services staff have experienced stress and poor mental health issues at work and are "more likely to experience a mental health problem than the general workforce but are less likely to take time off work as a result" (MIND 2019). Cases of bullying and harassment are regularly included in the official inspection reports and also covered in media.

Across the sector, the drive to professionalisation and modernisation of the workforce is showing positive results, but the pace and scale of changes varies across the three services. Moreover, as argued in the previous chapters, the changing complexity of the 999 demands has resulted into a period of unprecedented change (Wankhade 2018; Wankhade et al. 2015; Mansfield 2015; College of Policing 2015), in addition to the effects of Brexit, which is generating more anxiety and pressures for the emergency work force. In this chapter, we identify and discuss the specific challenges faced by the three main emergency services. We will offer our analysis and draw some conclusions based upon the current evidence.

This chapter is structured as follows. The firsts section draws more broadly on the health and well-being issues witnessed by the emergency services work force. We then dwell deeper on the issues of staff sickness and then discuss the growing cases of harassment and bullying in the workforce. We then discuss the blue light health and well-being frameworks being developed along with other support systems in place. This is followed by some concluding remarks.

Crisis of Anxiety: Health and Well-Being Issues in the Emergency Services

A diverse, modern and healthy workforce is 'sine qua non' for effective working of the emergency services around the world. Increasingly, the importance of workforce well-being and organisational resilience is being reported more frequently in the context of the three blue light services (Wankhade et al. 2018; Murphy and Greenhalgh 2018; Tehrani and Piper 2011; Williams et al. 2010; Brough 2005; Wagner et al. 2002). In the UK, huge budgetary cuts (NAO 2015a, b, 2017) accompanied by legislative changes brought through the Policing and Crime Act, 2017 to a more joint-up working between the three emergency services (HM Government 2016; Charman 2014) are putting more pressures on the emergency service workers, since these changes require the police, the paramedics and the firefighters to take over aspects of each other's roles/functions (Tehrani and Hesketh 2019).

The mainstream 'work intensification' literature is rather well developed in different organisational settings (Lyng 2005; Green 2004; Burke 2009; McFarlane and Bryant 2007; Felstead et al. 2013; Boxall and Macky, 2014; Piasna 2018; Gregg 2011), but the emergency services have received little attention so far. Management interest about emergency services is slow, but specific accounts of such 'extreme' work are now emerging (Granter et al. 2019; Wankhade et al. 2018; Turnbull and Wass 2015). Hewlett and Luce (2006) classify any work as 'extreme' which is in the excess of 60 hours per week and which involves high-intense and high-risk roles. Work done by the emergency services can be characterised as 'extreme' but comes at a cost of perverse consequences. For instance, Maguire et al. (2014) consider the work of ambulance paramedics as dangerous occupation. Granter et al. (2019), in a recent study, identified four distinct but overlapping dimensions of work intensity, namely temporal, emotional, organisational and physical, within the setting of the English ambulance services. In another study with the police, Turnbull and Wass (2015) found evidence of long working hours for the inspectors in almost every police force in the UK in excess of the maximum of 48 hours per week regulation. The study concluded that such long hours are "detriment to the workforce (adverse health effects), the organization (unrepresentative of the community) and wider society (degraded interactions)" (ibid., p. 525). In another study pertaining to the firefighters, Berninger et al. (2010) have also highlighted the increased risk of developing delayed onset PTSD when 'firefighters work within proximity to the disaster and when first responding to the incident'.

Cases of stress, mental illness, stress and post-traumatic stress disorder (PTSD) are on the rise in these organisations (Fraess-Phillips et al. 2017; Baydoun et al. 2016; Mohd Dahlan et al. 2010). The psychological impact of traumatic events on emergency service personnel is well documented in the literature (West and Murphy 2016; Donnelly and Bennett 2014). In a study focusing on the American and Canadian firefighters, PTSD rates among firefighters were shown to be greater than the population averages (Corneil et al. 1999). Several studies have reported cases of burn-out, divorce, drug abuse and suicide in emergency services workers (Wankhade 2018; Sterud et al. 2011; Dembe 2009; Wagner et al. 2002; van der Ploeg and Kleber 2003). Similarly Sterud et al. (2006) in their systematic review on the health status of the ambulance workers, identified a wide range of health problems such as post-traumatic stress symptoms, mental and somatic problems, injuries, fatal accidents and infectious diseases in the ambulance services staff. The frontline emergency services responders do not always get sufficient time (and warning) to prepare and deal with the operational and emotional demands of their jobs, and as a result their ability to cope up can become 'overwhelmed' (Tehrani and Hesketh 2019, p. 2; Tehrani and Piper 2011).

This discussion supports our argument that the psychological wellbeing of emergency service staff significantly affects their functionality and organisational productivity while also having a 'ripple' effect on families (Dunn et al. 2015, p. 17). A study of the fire and emergency services in Western Australia has concluded that trauma exposure, social support and coping style also significantly contributes to elevated rates of PTSD symptomatology (Skeffington et al. 2017). Similarly, a recent in-depth review of the work-related PTSD has also confirmed that professional first responders, such as firefighters, ambulance personnel and police officers, have a greater risk of being exposed to traumatic events through their daily work (Skogstad et al. 2013). Supporting workforce and dealing with staff issues are proving to be a neglected management priority given the operational focus of these services with limited evidence of staff engagement in the design of organisational systems (McCann et al. 2013, 2015; Wankhade et al. 2015, 2018). The next section details the issue of sickness absence in the emergency services.

SICKNESS ABSENCE IN POLICE

The problem of sickness absence is a wider one across the emergency services and pervades across the police services (Gerber et al. 2010; Cooper and Ingram 2004). For example there were 2362 full-time equivalent police officers on long-term sick leave in the 43 forces in England and Wales at the end of March 2018, accounting for 1.9% of police officers in England and Wales (see Fig. 5.1).

There is also variation in sickness levels across ranks, with higher levels of sickness absence reported for constables (2%) and sergeants (1.6%) as compared to chief inspector or above (1.2%). Please see Fig. 5.2.

Statistics further reveal that sickness rates have been consistently higher among females than among males, with 2.5% of female officers on long-term sick leave, compared with 1.7% of male officers (Home Office 2018, p. 39).

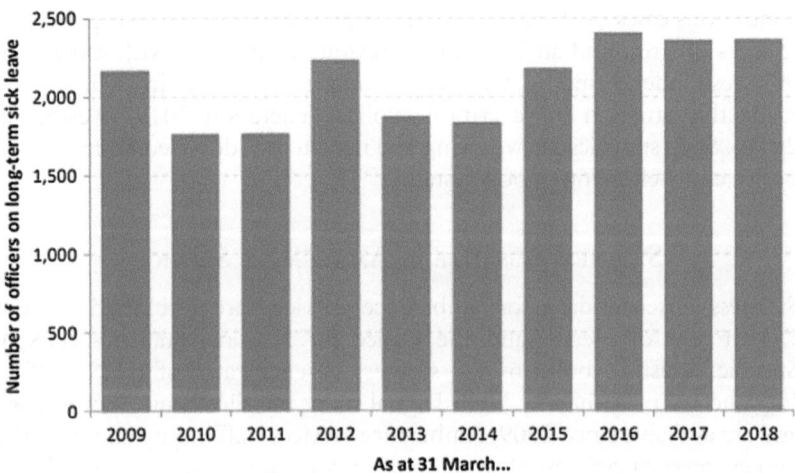

Fig. 5.1   Number of officers (FTE) on long-term sick leave, between 31 March 2009 and 31 March 2018, England and Wales. Source: Home Office (2018, p. 38)

Rank	Proportion on long-term sick leave
Constable	2.0%
Sergeant	1.6%
Inspector	1.5%
Chief Inspector or above	1.2%

Fig. 5.2   Proportion of officers Full Time Equivalent (FTE) on long-term sick leave, by rank, as of 31 March 2018, England and Wales. Source: Home Office (2018, p. 39)

Reviewing the sickness absence in the police, the Health and Safety Executive (HSE) (2007) concluded that current "approaches of the policies varied from the very punitive to stressing the forces' duty of care to staff. The most effective policies accepted that ill health is unavoidable, recognised the importance of individuals feeling valued and had measures in place to encourage and support return to work" (p. vi).

A recent survey by the Police Federation of England and Wales (PFEW) survey reported high levels of under-staffing in the forces which results in continuous cases of stress and PTSD. Almost 80% of officers experienced feelings of stress and anxiety in the previous 12 months, with more than 90% respondents stating that the traumatic experience in their careers made the situation more critical (Police Federation 2019; Westbrook 2019). Such statistics are worrying and need to be addressed to ensure the health and well-being of police staff.

SICKNESS ABSENCE IN AMBULANCE SERVICE

Sickness absence data for ambulance services are published by the Department of Health and the Office for National Statistics (ONS). Specific statistical bulletins for sickness absence rates of NHS staff in England are published by NHS Digital using the Electronic Staff Record and are available since 2009. Ambulance services staff in England have the highest level of sickness absence rates in comparison to other healthcare organisations within the National Health Service (NHS), showing an increase every year (Table 5.1).

As against the national average sickness absence rate of 4.20% over a seven-year period, ambulance trusts show an average absence rate of 5.78%, which is the highest for the entire NHS workforce included in the data set. The rates have shown a slight decline for the last two years, but more needs to be done to bring these high sickness rates down in relation to other NHS organisations. Reducing sickness absence has been recognised as central to the health and well-being of the staff (NHS 2019; NAO 2017; Association of Ambulance Chief Executives AACE 2017; Boorman 2009). A recent efficiency review (Carter 2016) expressed concern for the high level of sickness absence rates in the NHS and estimated that a 1% improvement in staff productivity will save the NHS £280 million a year, which equates to hospitals using new working methods that would save every member of staff five minutes on an eight-hour shift.

A further comparison of these sickness figures for different staffing groups in the NHS corroborates this disturbing scenario (see Table 5.2).

Against the overall average absence rate of 4.16% over the seven-year period, ambulance staff show an average absence rate of 6.2%, which is the highest amongst the entire NHS workforce included in the data set. We need to be cautious that these figures do not suggest a specific cau-

Table 5.1 Annual sickness absence rates by organisation type in the NHS

	2010–2011 (%)	2011–2012 (%)	2012–2013 (%)	2013–2014 (%)	2014–2015 (%)	2015–2016 (%)	2016–2017 (%)
England	**4.16**	**4.12**	**4.24**	**4.06**	**4.25**	**4.15**	**4.16**
Acute trusts	3.91	3.89	4.01	3.84	4.03	3.97	3.99
Ambulance trusts	**5.67**	**5.76**	**6.05**	**5.82**	**6.27**	**5.51**	**5.40**
Clinical Commissioning Groups	–	–	2.07	2.20	2.60	2.61	2.78
Commissioning Support Groups	–	–	–	2.69	3.05	2.82	2.84
Community Provider Trusts	4.64	4.60	4.65	4.47	4.65	4.57	4.66
Mental Health	4.95	4.89	4.94	4.74	4.88	4.78	4.79
Primary Care Trust (PCT)	4.20	3.93	3.09	3.26	2.15	–	–
Special Health Authority	3.69	3.47	3.56	3.30	3.47	3.29	3.17
SHA	2.31	2.13	2.55	–	–	–	–

Source: Adapted from NHS Sickness Absence Rates, Annual Summary Tables, 2010 to 2016–17, NHS Digital, 25 July 2018a (Table 5.2)

Table 5.2 Annual sickness absence rate by staff groups in the NHS

	2010–2011 (%)	2011–2012 (%)	2012–2013 (%)	2013–2014 (%)	2014–2015 (%)	2015–2016 (%)	2016–2017 (%)
Total	**4.16**	**4.12**	**4.24**	**4.06**	**4.25**	**4.15**	**4.16**
Professionally qualified clinical staff	3.65	3.61	4.24	4.06	3.70	3.56	3.55
All HCSC doctors	1.16	1.19	1.25	1.22	1.21	1.23	1.25
Total HCHS non-medical staff	4.46	4.42	4.56	4.37	4.58	4.47	4.49
Qualified nursing, mid-wifery and health well-being staff	4.59	4.55	4.73	4.50	4.64	4.49	4.48
Total qualified scientific, therapeutic and technical staff	2.91	2.88	2.98	2.87	3.01	2.92	2.98
Qualified ambulance staff	**6.18**	**6.18**	**6.55**	**6.20**	**6.85**	**5.86**	**5.49**
Support to clinical staff	5.48	5.41	5.54	5.32	5.61	5.50	5.55
NHS infrastructure support	3.68	3.66	3.75	3.58	3.74	3.74	3.73
Other non-medical staff or those with unknown classification	1.10	1.03	2.61	1.27	1.35	1.41	1.66

Source: Adapted from NHS Sickness Absence Rates, Annual Summary Tables, 2010 to 2016–17, NHS Digital, 25 July 2018b (Table 5.3)

sality on specific reasons for sickness absence within the NHS in England (Wankhade 2016). Such high sickness absence rates are a cause for concern, but many recent studies (Mishra et al. 2010; Cotton and Hart 2003; Hegg-Deloye et al. 2014) have argued that paramedics accumulate a set of risk factors, including acute and chronic stress, that can lead to cardiovascular diseases and post-traumatic disorders, with sleep problems and obesity being prevalent among paramedics in a wide range of settings around the world.

Sickness Absence in Fire Services

Sickness absence, also referred to as absence management in the fire and rescue services, is monitored locally by each organisation. Comparative data available for 2015–2016 suggest that sickness absence in on the rise (see Fig. 5.3). Overall, retained and non-uniformed staff categories constitute more than 95% of the staff sickness absence.

Comparative figures for 2015/2016 over 2014/2015 show a marked increase compared to the previous year's figures. For instance, overall staff sickness shows a rise of 2%, showing corresponding increase for the retained staff (5%) and the fire control staff (15%). On a positive side, sickness figures showed a decline for the full-time staff (−0.6%) and non-uniformed staff (−2.4%). This is detailed in Fig. 5.4.

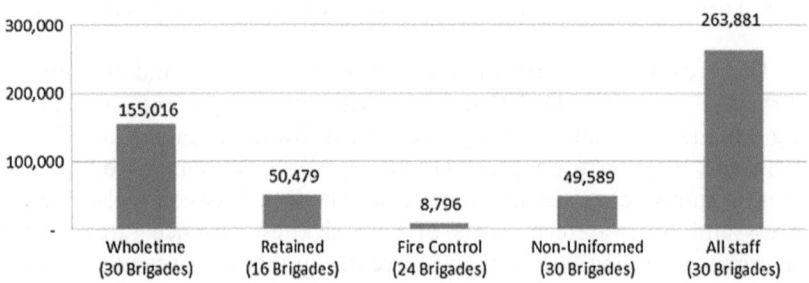

Fig. 5.3 Total days/shifts lost to sickness. Source: Cleveland Fire Brigade (2015, p. 2)

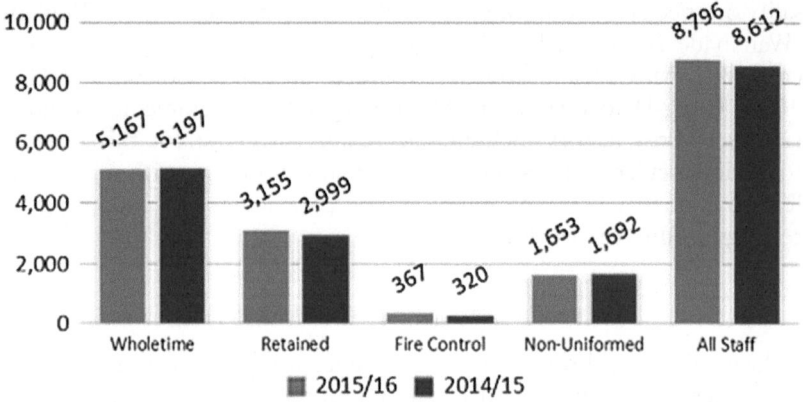

Fig. 5.4 Average days/shifts lost to sickness per brigade—2014/2015–2015/2016. Source: Cleveland Fire Brigade (2015, p. 2)

According to the report, the five key causes for all the categories of staff include:

- Musculoskeletal injuries (such as back, lower limb, shoulder and neck)
- Respiratory diseases (such as cold, chest and asthma)
- Gastrointestinal issues (such as abdominal pain, gastroenteritis, vomiting, diarrhoea)
- Mental health (such as stress, anxiety and depression)
- Hospital/Postoperative issues (such as cancer, virus and neurological)

These causes also find mention in a major health and well-being survey at work (CIPD 2018, p. 6), with more organisations reporting 'mental ill-health' among their most common causes for short-term and long-term absence. However, workload remains by far the most common cause of stress at work. The link between work-related stress and other determinants of high sickness absence rates in the emergency service workers is currently under-researched. Further empirical work will help to address this knowledge and research gap and the need for greater management and policy attention (CIPD 2018; Wankhade 2016).

WORKPLACE BULLYING AND HARASSMENT

There are multiple definitions of what constitutes 'unacceptable behaviour' (Barling et al. 2009), but it is generally understood as "behaviour by an individual or individuals within or outside an organisation that is intended to physically or psychologically harm a worker or workers and occurs in a work-related context" (Schat and Kelloway 2005, p. 191).

Workplace bullying is generally understood as a psychological and non-physical form of unacceptable behaviour (Sprigg et al. 2010; Rayner and Hoel 1997; Rayner 1997) and can be defined as an "escalating process in the course of which the person confronted ends up in an inferior position and becomes the target of systematic negative social acts" (Einarsen et al. 2003, p. 15). Under the Equality Act 2010, harassment is unlawful and is defined as "unwanted conduct related to a relevant protected characteristic, which has the purpose or effect of violating an individual's dignity or creating an intimidating, hostile, degrading, humiliating or offensive environment for that individual". However the words bullying and harassment are often used interchangeably (Arbitration and Conciliation Service, ACAS 2014).

The problem of bullying and harassment is particularly serious within the UK healthcare workers including the ambulance staff. Ambulance staff continue to face high levels of discrimination from their peer, managers and the public. In a study conducted by the King's Fund (2015), the 2014 NHS Staff Survey data was analysed for reported discrimination within the NHS, between managers and staff, between colleagues, but also between patients and members of the public on grounds of age, gender, religion, sexual orientation, disability and ethnicity. This study measured discrimination reported by NHS staff on the aforementioned dimensions (see Table 5.3).

The findings make an uncomfortable reading. There was a wide variation for the reported levels of discrimination based upon the type of the trust and the chosen variables, but ambulance trusts showed the highest levels of reported discrimination. More worryingly, as the table shows, reported levels of discrimination reported by ambulance staff are highest for eight of the nine dimensions in the study, except on grounds of ethnicity. The gender bias is reversed in the case of ambulance trusts, and more women reported discrimination than men as compared to any other type of organisations. Discrimination from

Table 5.3 Reported discrimination—difference by NHS Trust type

		Overall (%)	Acute (%)	Community (%)	MH/ LD (%)	Ambulance (%)	Other (%)
	Any discrimination	**11.9**	11.7	8.9	12.9	**19.7**	5.3
Discrimination from	Patients, relatives, public	**5.9**	5.6	3.7	7.1	**10.9**	1.5
	Manager, team leader, other colleagues	**8.0**	8.1	6.3	7.7	**12.6**	4.3
Discrimination on the basis of	Ethnic background	**4.3**	4.5	2.3	4.8	**3.0**	0.8
	Gender	**2.2**	2.0	1.6	2.7	**6.1**	1.1
	Religion	**0.6**	0.6	0.3	0.7	**0.8**	0.1
	Sexual orientation	**0.6**	0.5	0.3	0.8	**2.1**	0.2
	Disability	**0.9**	0.8	0.9	1.1	**1.4**	0.6
	Age	**2.2**	2.1	1.5	2.5	**5.4**	1.2

Source: Adapted from The King's Fund (2015, p. 14)

managers and colleagues was also the highest in the case of ambulance trusts (The King's Fund 2015).

These challenges continued to be reflected in the subsequent surveys. The 2016 surveys reported again highlighted low staff morale, with some fundamental problems with staffing, morale, pay and management in the ambulance trusts (Appleby and Dayan 2017). The latest surveys in 2017 continue to show that "bullying and harassment remains a pervasive problem in the health sector with 24 per cent of all NHS staff (one in four people) having reported that they have experienced bullying in some way" (NHS Employers 2018). The responses from the ambulance staff surveys paint a very 'damning' picture:

> Compared with other types of trust, as well as clinical commissioning groups (CCGs), their training and development is strikingly poor, they are far worse than other NHS organisations for discrimination and equal opportunities, worst for illness due to work-related stress, worst for organisational and management interest in their health and wellbeing, worst at giving staff appraisals, worst for appraisal quality, worst for team working and worst for staff engagement. (Vize 2018)

There are clearly issues around the management and performance regime currently used by the ambulance services. A range of unintended consequences of ambulance response time targets have been documented (Wankhade 2011). Recent evidence suggests that ambulance work is continuing to be quite intense which is putting further pressure on staff and their well-being (Granter et al. 2019; McCann et al. 2015). Performance targets continue to focus the minds of staff, while still influencing the process of change (Wankhade et al. 2018; Heath et al. 2018; Wankhade 2018).

Bullying is now reported in official publications. Some of the recent Care Quality Commission (CQC) inspection reports have brought to fore the issues of bullying and harassment. Its report for the South East Coast Ambulance Service (CQC 2018) has highlighted continuing perceptions of bullying and harassment in some areas across the trust and reported that "some members of staff were still affected by previous concerns surrounding bullying and harassment" (CQC 2018, p. 7). An independent Bullying and Harassment Review (Twist 2017) found a breakdown of communication between the management and union staff in another ambulance trust and concluded that the trust "appears to tolerate a bullying culture and managers adopt, in the main, a bullish approach to people management with a culture of 'face fits' and nepotism so 'friends' are put into positions of power rather than those who may be better suited to management" (p. 4). These are not isolated cases and continue to happen after the much publicised case of the London Ambulance Service in 2015, wherein bullying and harassment along with a culture of fear were identified as major causes for concern (CQC 2015).

Such cases are not unique to the ambulance services alone. The fire and rescue services were criticised by the UK Home Secretary for "lack of diversity and allowing a culture of 'bullying and harassment' to flourish in some parts of England and Wales and warned there was 'no excuse' for the 'toxic and corrosive' attitudes identified in some of the country's fire and rescue services" (Drury 2016). The Policing and Crime Act 2017 has set out that the newly created Her Majesty's Inspectorate of Constabulary and Fire and Rescue Services (HMICFRS) will be inspecting the fire and rescue authorities in England with a focus on three aspects—efficiency, effectiveness and leadership/care for people. In the first instance, 14 fire and rescue services had been inspected by the end of 2018. On the issues of ensuring fairness, promoting diversity and promoting the right values and culture, one service was rated as inadequate, nine were rated as needing improvements and only four were rated as good (see HMICFRS 2019 for further details).

Cases of bullying and harassment are being dealt by the police services too, and instances of 'macho, arrogant, bullying culture' in which whistle-blowing has never been fully embedded have been reported (Williams 2015). The College of Policing in a highly critical report on the miscon-duct by chief officers has highlighted a 'bullying' culture that prevents junior staff from confronting their seniors over misconduct (Hales et al. 2015). Another report by the College of Policing makes a case for the role of senior leaders in creating ethical organisations to counter the percep-tion that "blowing the whistle" can be detrimental to a person's career (Quinton 2015, p. 2).

BLUE LIGHT WELL-BEING INITIATIVES

In recognition of the growing crisis around workforce health and well-being, efforts are being made to provide more structured support. The College of Policing has developed a *Blue Light Wellbeing Framework* (2017) in partnership with other agencies (such as Public Health England) to provide advice and support to police staff and create a positive working environment. The framework applies a self-scoring inventory on a series of questions on the following domains:

1. Leadership
2. Absence management
3. Creating the environment
4. Mental health
5. Protecting the workforce
6. Personal resilience

A series of questions guide the level of engagement within the organisa-tion and what needs to be done. This will allow the support systems in place to be benchmarked against an independent set of standards that have been tailored to meet the specialist needs of emergency services staff. The newly created *National Police Wellbeing Service* is coordinating with the well-being support to staff across the police forces using the framework as detailed above.

Consensus statements for each of the three emergency services to work together with other organisations to help improve public health and well-being are being drawn (Public Health England 2015, 2018; Association of Ambulance Chief Executives AACE 2017). Such an approach involving public health along with targeted interventions can be quite useful. In a recent study in an Australian Fire and Rescue Service, it was demonstrated that a four-hour face-to-face mental health training programme "could lead to a significant reduction in work-related sickness absence, with an associated return on investment of £9.98 for each pound spent on such training" (Milligan-Saville et al. 2017, p. 850). Further research can help to test these findings in other settings but the study offers a potential of economic benefits of a brief workplace intervention to the emergency services.

Conclusion

As argued in the previous chapter, the emergency services are witnessing a period of unprecedented change amidst budgetary pressures and workforce shortages and sickness. It is somewhat forgotten how people at various levels are managing to cope up with all the changes and uncertainty or whether they are just managing to survive. These are not only moral issues relating to staff health and well-being, but also of major organisational significance in terms of performance, employee engagement, cost of ill health and potential litigation claims.

We need to do more to support our staff to cope up with work pressure. There is quite some evidence to support this hypothesis. According to the World Health Organisation, stress is the 'Health Epidemic of the 21st Century' estimated to be costing American businesses alone $300 billion a year (Finch 2016). For each year over the last five years, work-related stress, depression or anxiety has been the single most reported complaint to the Health and Safety Executive HSE (2008). Liddell (2013) estimates the cost of this to police forces in England alone to be just under £100 million per year. In the HSE Report (2007, p. 94) on police sickness, "work was perceived to be a contributory factor to both short and long-term sickness".

If a greater understanding of resilience and the relationship between resilience and stress can be achieved, we will be better able to proactively implement and support interventions to assist employees to cope more effectively with the stress that is inherent in today's workplaces. There has never been a greater need to support the workforce in preparing them for these very uncertain and challenging times (Cooper and Hesketh 2017).

Research suggests that it is particularly important to have visible and sustained top management support for positive diversity and inclusion of policies and practices. But it is equally important that these are seen to be implemented effectively and consistently and are reinforced by middle management and frontline supervisors. We have also separately explored the relationship between cultures, performance measures and organisational change to understand how organisational culture is perpetuated and found the targets to be a significant factor impeding the process of change (Wankhade et al. 2018; see also Granter et al. 2019). More empirical work will shed additional insights on these important issues.

The next chapter deals with the interesting notion of organisational culture(s) in the emergency services in the light of the changing professional identities of the blue light workforce.

REFERENCES

Appleby, J., & Dayan, M. (2017). *Nuffield Winter Insight Briefing 3: The ambulance service*. London: Nuffield Trust. Retrieved October 27, 2018, from https://www.nuffieldtrust.org.uk/files/2017-04/winter-pressures-ambulances-briefing-web-final.pdf.

Arbitration and Conciliation Service, ACAS. (2014). *Bullying and harassment at work: A guide for employees*. London: ACAS. Retrieved December 10, 2018, from http://www.acas.org.uk/media/pdf/r/l/Bullying-and-harassment-at-work-a-guide-for-employees.pdf.

Association of Ambulance Chief Executives AACE. (2017). *Working together with ambulance services to improve public health and wellbeing*. Launch of Consensus Statement. London: AACE. Retrieved July 10, 2018, from https://aace.org.uk/wp-content/uploads/2017/02/Launch-of-PH-Consensus-Statement-ALF-20174PP-V1.pdf.

Barling, J., Dupré, K. E., & Kelloway, E. K. (2009). Predicting workplace aggression and violence. *The Annual Review of Psychology, 60*, 671–692.

Baydoun, M., Dumit, N., & Daouk-Öyry, L. (2016). What do nurse managers say about nurses' sickness absenteeism? A new perspective. *Journal of Nursing Management, 24*(1), 97–104.

Berninger, A., Webber, M. P., Niles, J. K., Gustave, J., Lee, R., Cohen, H. W., & Prezant, D. J. (2010). Longitudinal study of probable post-traumatic stress disorder in firefighters exposed to the World Trade Center disaster. *American Journal of Industrial Medicine, 53*(12), 1177–1185. https://doi.org/10.1002/ajim.20894.

Boorman, S. (2009). *NHS health and well-being review: Interim and final report*. London: Department of Health.

Boxall, P., & Macky, K. (2014). High-involvement work processes, work intensification and employee well-being. *Work, Employment & Society, 28*(6), 963–984.

Brough, P. (2005). A comparative investigation of the predictors of work-related psychological well-being within police, fire and ambulance workers. *New Zealand Journal of Psychology, 34*(2), 127–134.

Burke, R. J. (2009). Working to live or living to work: Should individuals and organizations care? *Journal of Business Ethics, 84*(S2), 167–172.

Care Quality Commission. (2015). *London ambulance service NHS trust: Quality report.* London: CQC.

Care Quality Commission CQC. (2018). *Inspection report-south east coast ambulance service NHS Foundation Trust.* London: CQC.

Carter, L. (2016). *Operational productivity and performance in English NHS acute hospitals: Unwarranted variations: An independent report for the Department of Health by Lord Carter of Coles.* London: Her Majesty's Government.

Charman, S. (2014). Blue light communities: Cultural interoperability and shared learning between ambulance staff and police officers in emergency response. *Policing and Society, 24*(1), 102–119.

Chartered Institute of Personnel and Development CIPD. (2018, May). *Health and well-being at work* (Survey report). London: CIPD. Retrieved December 20, 2018, from https://www.cipd.co.uk/knowledge/culture/well-being/health-well-being-work.

Cleveland Fire Brigade. (2015). National fire & rescue service: Occupational health performance report, April 2015–March 2016.

College of Policing. (2015). College of policing analysis: Estimating demand on the police service. Coventry: College of Policing. Retrieved November 10, 2018, from https://www.college.police.uk/News/College-news/Documents/Demand%20Report%2023_1_15_noBleed.pdf.

College of Policing. (2017). *Blue light wellbeing framework: Organisational development and international faculty.* Coventry: College of Policing. Retrieved November 17, 2018, from https://www.college.police.uk/What-we-do/Support/Health-safety/Documents/Blue-Light-Wellbeing-Framework.pdf.

Cooper, C. L., & Hesketh, I. (2017). *Managing health and wellbeing in the public sector a guide to best practice* (1st ed.). London: Routledge.

Cooper, C., & Ingram, S. (2004). *Retention of police officers: A study of resignations and transfers in ten forces.* RDS Occasional Paper No. 86. London: Home Office.

Corneil, W., Beaton, R., Murphy, S., Johnson, C., & Pike, K. (1999). Exposure to traumatic incidents and prevalence of posttraumatic stress symptomatology in urban firefighters in two countries. *Journal of Occupational Health Psychology, 4*(2), 131–141. https://doi.org/10.1037/1076-8998.4.2.131.

Cotton, P., & Hart, P. M. (2003). Occupational wellbeing and performance: A review of organisational health research. *Australian Psychologist, 38*(2), 118–127.

Dembe, A. E. (2009). Ethical issues relating to the health effects of long working hours. *Journal of Business Ethics, 84*(S2), 195–208.

Donnelly, E. A., & Bennett, M. (2014). Development of a critical incident stress inventory for the emergency medical services. *Traumatology: An International Journal, 20*(1), 1–8.

Drury, I. (2016, May 25). Theresa May slams fire service chiefs for allowing 'bullying and harassment' to flourish as she unveils sweeping reforms. *Mail Online.*

Retrieved November 8, 2018, from https://www.dailymail.co.uk/news/article-3609247/Theresa-slams-fire-service-chiefs-allowing-bullying-harassment-flourish-unveils-sweeping-reforms.html.

Dunn, R., Brookes, S., Rubin, J., & Greenberg, N. (2015). Psychological impact of traumatic events: Guidance for trauma-exposed organisations. *Occupational Health at Work, 12*(1), 17–21.

Einarsen, S., Hoel, H., Zapf, D., & Cooper, C. L. (2003). The concept of bullying at work: The European tradition. In S. Einarsen, H. Hoel, D. Zapf, & C. L. Cooper (Eds.), *Bullying and emotional abuse in the workplace: International perspectives in research and practice* (pp. 3–30). London: Taylor and Francis.

Felstead, A., Gallie, D., Green, F., & Inanc, H. (2013). *Work intensification in Britain: First findings from the skills and employment survey 2012*. London: Institute of Education.

Finch, G. (2016). *Stress: The health epidemic of the 21st century*. Elsevier: SciTech Connect, April 26, 2016. Retrieved February 10, 2019, from http://scitech-connect.elsevier.com/stress-health-epidemic-21st-century/.

Fraess-Phillips, A., Wagner, S., & Harris, R. L. (2017). Firefighters and traumatic stress: A review. *International Journal of Emergency Services, 6*(1), 67–80.

Gerber, M., Hartmann, T., Brand, S., Holsboer-Trachsler, E., & Pühsea, U. (2010). The relationship between shift work, perceived stress, sleep and health in Swiss police officers. *Journal of Criminal Justice, 38*(6), 1167–1175.

Granter, E., Wankhade, P., McCann, L., Hyde, P., & Hassard, J. (2019). Multiple dimensions of work intensity: Ambulance work as edgework. *Work Employment and Society, 33*(2), 280–297.

Green, F. (2004). Why has work effort become more intense? *Industrial Relations, 43*(4), 709–741.

Gregg, M. (2011). *Work's Intimacy*. Cambridge: Polity Press.

Hales, G., May, T., Belur, J., & Hough, M. (2015). *Chief officer misconduct in policing: An exploratory study*. Ryton-on-Dunsmore: College of Policing, College of Police.

Health and Safety Executive. (2007). *Managing sickness absence in the police service: A review of current practices*. RR582 Research Report. HSE: London.

Health and Safety Executive HSE. (2008). *Health and safety statistics 2008/9*. HSE: Suffolk.

Heath, G., Radcliffe, J., & Wankhade, P. (2018). Performance management in the public sector: The case of the English ambulance service. In E. Harris (Ed.), *The Routledge companion to performance management and control* (pp. 417–438). London: Routledge.

Hegg-Deloye, S., Brassard, P., Jauvin, N., et al. (2014). Current state of knowledge of post-traumatic stress, sleeping problems, obesity and cardiovascular disease in paramedics. *Emergency Medicine Journal, 31*, 242–247.

Hewlett, S. A., & Luce, C. B. (2006). Extreme jobs: The dangerous allure of the 70-hour workweek. *Harvard Business Review, 84*(12), 49–59.

HM Government. (2016). *Enabling closer working between the emergency services summary of consultation responses and next steps.* Retrieved February 18, 2018, from https://assets.publishing.service.gov.uk/government/uploads/system/uploads/attachment_data/file/495371/6.1722_HO_Enabling_Closer_Working_Between_the_Emergency_Services_Consult....pdf.

HMICFRS. (2019). Fire and rescue service inspections 2018/19 – summary of findings from tranche 1. Retrieved February 19, 2019, from https://www.justiceinspectorates.gov.uk/hmicfrs/publications/fire-and-rescue-service-inspections-2018-19/.

Home Office. (2018). Police workforce, England and Wales, 31 March 2018. *Statistical Bulletin, 11/18* (London: Crime and Policing Statistics, Crime and Policing Analysis Unit).

Liddell, E. (2013). *Prevalence on PTSD, compassion fatigue and burnout in the emergency services.* Presentation to the Police Federation Health and Safety Conference 30th September. Leatherhead: Police Federation Headquarters.

Lyng, S. (Ed.). (2005). *Edgework: The sociology of risk-taking.* London: Routledge.

Maguire, B. J., O'Meara, P. F., Brightwell, R. F., O'Neill, B. J., & Fitzgerald, G. J. (2014). Occupational injury risk among Australian paramedics: An analysis of national data. *The Medical Journal of Australia, 200*(8), 477–480.

Mansfield, C. (2015). *Fire works: A collaborative way forward for the fire and rescue service.* London: New Local Government Network (NLGN).

McCann, L., Granter, E., Hyde, P., & Hassard, J. (2013). Still Blue-Collar after all these years? An ethnography of the professionalization of emergency ambulance work. *Journal of Management Studies, 50*(5), 750–776.

McCann, L., Hassard, J., Granter, E., & Hyde, P. (2015). 'You can't do both: something will give': Limitations of the targets culture in managing UK health care workforces. *Human Resource Management, 54*(5), 773–791.

McFarlane, A. S., & Bryant, R. A. (2007). Post-traumatic stress disorder in occupational settings: Anticipating and managing the risk. *Occupational Medicine, 57*(6), 404–410.

Milligan-Saville, J. S., Tan, L., Gayed, A., Barnes, C., et al. (2017). Workplace mental health training for managers and its effect on sick leave in employees: A cluster randomised controlled trial. *Lancet Psychiatry, 4,* 850–858.

MIND. (2019). *Bluelight mental health info.* Retrieved December 17, 2018, from https://www.mind.org.uk/information-support/working-in-the-emergency-services/.

Mishra, S., Goebart, D., Char, E., Dukes, P., & Ahmed, I. (2010). Trauma exposure and symptoms of post-traumatic stress disorder in emergency medical services personnel in Hawaii. *Emergency Medicine Journal, 27,* 708–711.

Mohd Dahlan, A. M., Mearns, K., & Flin, R. (2010). Stress and psychological well-being in UK and Malaysian fire fighters. *Cross Cultural Management: An International Journal, 17*(1), 50–61.

Murphy, P., & Greenhalgh, G. (Eds.). (2018). *Fire and rescue services: Leadership and management perspectives*. Cham: Springer.

National Audit Office. (2015a). *Financial sustainability of police forces in England and Wales*. Retrieved February 18, 2018, from www.nao.org.uk/wp-content/uploads/2015/06/Financial-sustainability-of-policeforces.pdf.

National Audit Office. (2015b). *Impact of funding reductions on fire and rescue services*. Retrieved February 18, 2018, from www.nao.org.uk/report/impact-of-funding-reductions-on-fire-and-rescue-services/.

National Audit Office. (2017). *NHS ambulance services*. Retrieved February 18, 2018, from www.nao.org.uk/wp-content/uploads/2017/01/NHS-Ambulance-Services.pdf.

NHS. (2019). *NHS long term plan*. London: NHS England.

NHS Digital. (2018a). NHS sickness absence rates, annual summary tables, 2010–11 to 2016–17. *NHS Digital*, July 2017 (Table 2).

NHS Digital. (2018b). NHS sickness absence rates, annual summary tables, 2010–11 to 2016–17. *NHS Digital*, July 2017 (Table 3).

NHS Employers. (2018). Tackling bullying in the NHS. *NHS Employers*. Retrieved December 19, 2018, from https://www.nhsemployers.org/bullyingand-harassment.

Piasna, A. (2018). Scheduled to work hard: The relationship between non-standard working hours and work intensity among European workers (2005–2015). *Human Resource Management Journal, 28*(1), 167–181.

Police Federation. (2019, February 13). Government must face facts – Extreme stress in policing is real. *Police Federation*. Retrieved February 17, 2019, from http://www.polfed.org/newsroom/6930.aspx.

Public Health England. (2015). *Consensus statement on improving health and well-being between NHS England*. Public Health England, Local Government Association Chief Fire Officers Association and Age UK. Retrieved November 10, 2018, from https://www.england.nhs.uk/wp-content/uploads/2015/09/joint-consens-statmnt.pdf.

Public Health England. (2018). *Policing, health and social care consensus: Working together to protect and prevent harm to vulnerable people*. Retrieved July 20, 2018, from https://www.npcc.police.uk/Publication/NEW%20Policing%20Health%20and%20Social%20Care%20consensus%202018.pdf.

Quinton, P. (2015). *Police leadership and integrity: Implications from a research programme*. Ryton-on-Dunsmore: College of Policing, College of Police.

Rayner, C. (1997). The incidence of workplace bullying. *Journal of Community and Applied Social Psychology, 7*, 199–208.

Rayner, C., & Hoel, H. (1997). A summary review of literature relating to workplace bullying. *Journal of Community and Applied Social Psychology, 7*, 181–191.

Schat, A., & Kelloway, E. K. (2005). Workplace violence. In J. Barling, E. K. Kelloway, & M. R. Frone (Eds.), *Handbook of workplace stress* (pp. 189–218). Thousand Oaks, CA: Sage.

Skeffington, P. M., Rees, C. S., & Mazzucchelli. (2017). Trauma exposure and post-traumatic stress disorder within fire and emergency services in Western Australia. *Australian Journal of Psychology, 69*(1), 20–28.

Skogstad, M., Slorstad, M., Lie, A., Conradi, H. S., Heir, T., & Weisaeth, L. (2013). Work-related posttraumatic stress disorder. *Occupational Medicine, 63,* 175–182.

Sprigg, C. A., Martin, A., Niven, K., & Armitage, C. J. (2010). *Unacceptable behaviour, health and wellbeing at work A cross-lagged longitudinal study.* Report submitted to the IOSH Research Committee. Wigston: Institute of Occupational Safety and Health IOSH, IOSH. Retrieved October 15, 2018, from https://www.iosh.co.uk/bullying.

Sterud, T., Ekeberg, O., & Hem, E. (2006). Health status in the ambulance services: A systematic review. *BMC Health Services Research, 6,* 82.

Sterud, T., Hem, E., Lau, B., & Ekeberg, O. (2011). A comparison of general and ambulance specific stressors: Predictors of job satisfaction and health problems in a nationwide one-year follow-up study of Norwegian ambulance personnel. *Journal of Occupational Medicine and Toxicology, 6,* 10.

Tehrani, N., & Hesketh, I. (2019). The role of psychological screening for emergency service responders. *International Journal of Emergency Services, 8*(1), 4–19.

Tehrani, N., & Piper, N. (2011). Traumatic stress in the police service. In N. Tehrani (Ed.), *Managing trauma in the workplace: Supporting workers and organisations* (pp. 17–32). London: Routledge.

The King's Fund. (2015). *Making the difference: Diversity and inclusion in the NHS.* London: The King's Fund.

Turnbull, P., & Wass, V. (2015). Normalizing extreme work in the police service? Austerity and the inspecting ranks. *Organization, 22*(4), 512–529.

Twist, A. (2017). *East of England ambulance service NHS trust-independent bullying and harassment review.* The Andrea Adams Consultancy.

Van Der Ploeg, E., & Kleber, R. J. (2003). Acute and chronic job stressors among ambulance personnel: Predictors of health symptoms. *Occupational Environment Medicine, 60,* 40–46.

Vize, R. (2018, March). *NHS survey reveals staff are determined to make the best of tough conditions. The Guardian Online.* Retrieved June 17, 2018, from https://www.theguardian.com/healthcare-network/2018/mar/09/nhs-survey-staff-determined-best-tough-conditions.

Wagner, D., Heinrichs, M., & Ehlert, U. (2002). Prevalence of symptoms of post-traumatic stress disorder in German professional firefighters. *American Journal of Psychiatry, 155,* 1727–1732.

Wankhade, P. (2011). Performance measurement and the UK emergency ambulance service. *International Journal of Public Sector Management, 24*(5), 384–402.

Wankhade, P. (2016). Staff perceptions and changing role of pre-hospital profession in the UK ambulance services: An exploratory study. *International Journal of Emergency Services, 5*(2), 126–144.

Wankhade, P. (2018). The crisis in NHS ambulance services in the UK: Let's deal with the 'elephants in the room'! *Ambulance Today, 15*(1), 13–17.

Wankhade, P., Heath, G., & Radcliffe, J. (2018). Cultural change and perpetuation in organisations: Evidence from an English Emergency Ambulance Service. *Public Management Review, 20*(6), 923–948.

Wankhade, P., Radcliffe, J., & Heath, G. (2015). Organizational and professional cultures: An ambulance perspective. In P. Wankhade & K. Mackway-Jones (Eds.), *Ambulance services: Leadership and management perspectives* (pp. 65–80). New York: Springer.

West, D., & Murphy, P. (2016). Managerial and leadership implications of the retained duty system in English fire and rescue services: An exploratory study. *International Journal of Emergency Services, 5*(2). ISSN: 2047-0894.

Westbrook, I. (2019, February 13). Police shortages: 'Working alone left me with PTSD'. *BBC News.* Retrieved February 17, 2019, from https://www.bbc.co.uk/news/uk-47212662.

Williams, M. (2015, March 28). Police bullying culture deterring whistleblowers, report warns. *The Guardian.* Retrieved June 14, 2018, from https://www.theguardian.com/uk-news/2015/mar/28/police-bullying-culture-detering-whistleblowers-report-warns.

Williams, V., Ciarrochi, J., & Deane, F. P. (2010). On being mindful, emotionally aware, and more resilient: Longitudinal pilot study of police recruits. *Australian Psychologist, 45*(4), 274–282.

CHAPTER 6

Professional Cultures and Changing Identities in the Emergency Services

Abstract This chapter concerns the role of professional culture(s) in the emergency services while these organisations embark on a professionalisation and modernisation agenda. The concept of organisational culture is enduring, and cultural change is now a popular prescription for organisational change since structural changes are not deemed adequate to bring real transformational change. The chapter first provides a conceptual understanding of organisational culture and change. In analysing these issues within the context of the emergency services, we review the evidence about the changing perceptions of the organisational and occupational culture(s) in the police, ambulance and fire services which historically has been quite negative. The conceptual complexity adds to the difficulties of bringing effective transformation. While the pace of such changes is slow, we see positive signs, most noticeable in the recruitment of more younger, diverse workforce which necessitates a more open mindset to understand and review the notion of organisational culture(s).

Keywords Ambulance • Police • Fire service • Culture and change • Changing identity • Reform • Roles • Professionalism

© The Author(s) 2020
P. Wankhade, S. Patnaik, *Collaboration and Governance in the Emergency Services*,
https://doi.org/10.1007/978-3-030-21329-9_6

Introduction and Background

The issue of organisational and professional culture(s) is now gaining importance for understanding the change management approaches in different organisational settings (Hatch 1993, 2000; Sackman 1992; Martin 1992; Smircich 1983; Ferlie and Shortell 2001). It has been argued that structural changes are not sufficient on their own to bring a desired change, and interpreting and changing organisational culture has some bearing on organisational performance and quality and is becoming a popular prescription for organisational reform (Jorritsma and Wilderom 2012; Konteh et al. 2010; Mannion et al. 2005).

The notion of professional culture(s) within the emergency services settings is also gaining academic attention. Many such accounts within the ambulance services (see Wankhade 2012; Wankhade and Brinkman 2014; Wankhade et al. 2018; McCann et al. 2013); the police services (see Loftus 2009; Charman 2013, 2014); and the fire and rescue services (Murphy and Greenhalgh 2018; Mansfield 2015) are steadily emerging. Within such accounts, a command and control culture, accompanied by a tendency to blame, hierarchical and top-down management style, resistance to change and being risk-averse are some of the historical factors associated with these organisations (Wankhade and Brinkman 2014; Lister 2014; Wankhade 2012). Loftus (2009) raises the issues of equality and diversity in policing, with an overwhelming white, male and heterosexual police workforce posing significant challenges for ethnic minorities, females and gay and lesbian members of the force or for those who aspire to be one.

Several high-profile policy reports on the state of affairs in the emergency services have highlighted the challenging landscape and changing identities of the workforce and service delivery. These include, amongst others, Taking Healthcare to the Patient 2 (Association of Ambulance Chief Executives 2011), College of Policing Leadership Review (2015), Winsor Police Pay and Conditions Review (2012), the Knight Fire Services Reform report (2013) and detailed analysis done by the National Audit Office (see NAO (2011, 2017) for ambulance details; NAO (2015a) for police services and NAO (2015b) for fire and rescue services). They all have highlighted a need for these organisations to create and maintain the right culture to deliver high-quality care that is responsive to peoples' needs and preferences. Need for cultural change underpins many of the recommendations in these reports.

This chapter critically analyses the changing nature of the professional cultures accompanied by changes in demand and service delivery in the three main emergency services and will document issues, challenges and opportunities, while analysing the implications of some of the recent tragic events such as the Grenfell Fire tragedy in London and the Manchester Arena bombing. The chapter is structured as follows. In the first section, we critically review the literature around organisational culture and change. We then discuss the current thinking around professional cultures in the emergency services and its implications for the modernisation and professionalisation agenda. This is followed by some concluding remarks.

ORGANISATIONAL CULTURE AND CHANGE

The notion of organisational culture is interesting but controversial due to contested positions on its interpretation. Some commentators view it as a 'variable', something which an organisation *has* (Konteh et al. 2010; see Peters and Waterman 1982; Deal and Kennedy 1982), whereas the counter position is to view it as a 'root metaphor', what the organisation *is* (Morgan 1986; Meek 1988). Many scholars have argued a pluralist view of culture (Martin 1992), but multiple definitions and viewpoints also exist in the literature.

Notwithstanding the multiple definitions and competing interpretations of what culture means (Browaeys and Price 2008), the notion that it is socially constructed and is often expressed in terms of patterns of behaviour is quite popular. Schein (2004, p. 17) defines culture as "pattern of shared basic assumptions that was learned by a group as it solved its problem of external adaptation and internal integration; and that has worked well enough to be considered valid, and therefore to be taught to new members as the correct way to perceive, think, and feel in relation to those problems". Schein (1985) also argues that culture is manifested at three distinct levels in an organisation:

1. 'Artefacts' or visible manifestations of culture
2. Espoused 'values and beliefs' at the second level
3. More deep-rooted 'assumptions' of real and unconscious beliefs and expectations.

In case of the emergency services, 'artefacts' could be seen as the green uniform for the ambulance paramedics, 'values and beliefs' can be the speedy response to save lives and 'assumptions' could constitute what is the core philosophy of each of the three services. The understanding of organisational culture(s) receives greater significance for such important public services which are currently confronted with reduced public spending, changes in legislation (as highlighted earlier in Chap. 1) and drive towards more specialist roles in a rapidly changing emergency service landscape (Woollard 2006, 2009). This has resulted in a new thinking of 'doing more with less', resulting in situations in which private bus operators are replacing traditional roles carried out by the ambulance services in providing non-emergency transport (Fitzgerald 2015) or back office functions being awarded to private contractors to save costs (BBC 2012). As a result, such "continuing expansion of new specialisms, seeking formal recognition of their professional status, has further increased the challenge presented by competing sub-cultures within the public sector as a whole" (Wankhade et al. 2018, pp. 925–926; Marsh 2018).

Additionally, the role of organisational sub-cultures is also interesting (Wankhade 2012; see Martin 1992). In a study of the Australian nurses in different hospital settings, it was found that "innovative and supportive subcultures have a clear positive relationship, while bureaucratic subcultures have a negative relationship" (Lok et al. 2005, p. 494). Another study (Carlström and Lars-Eric 2014) in a Swedish hospital setting concluded that the dominance of positive characteristics such as flexibility, cohesion and trust seemed to decrease change-resistance behaviour. Waddington (1999, p. 287), in his much-cited study on police canteen sub-cultures, concluded that occurrences in the canteen give purpose and meaning to the "inherently problematic occupational experience".

Analysis of cultural change follows the growing interest in the study of organisational culture. It has been argued that cultural change programmes can evoke differing emotional display, response and compliance to rules, and culture change initiatives should be "viewed, conceptualised and modelled as a continuum rather than as a dichotomous event" (Ogbonna and Harris 2002, p. 48). However, such growing interest in studying organisational culture is not accompanied by a strong evidence base on complexities and implications of cultural interventions in different settings, which remain a fascinating but under-researched phenomenon (Wankhade and Brinkman 2014; Diefenbach 2007; Bititci et al. 2007; Harris and Ogbonna 1998; Dingwall and Strangleman 2005; Dutton et al. 2001). However as highlighted earlier in the chapter, effecting culture change is high on the government agenda encompassing various elements of organisational and structural change. Meaningful gains from such culture change initiatives can be severely restricted without a clear understanding of any perverse behaviour or unintended consequences of a culture change programme (Ackroyd and Crowdy 1990; Legge 1994). It has also been argued that weak empirical relationship between performance and culture could be attributed to a lack of understanding about unintended consequences of any culture change programme (Wankhade and Brinkman 2014). The next section analyses the professional cultures and changing identities of the emergency services workforce in the three blue light services.

PROFESSIONAL CULTURES IN THE EMERGENCY SERVICES

Historically, cultures in the emergency services have been treated as monolithic, exhibiting characteristics such as authoritarianism, top-down, hierarchical command and control and resistant to change (Wankhade and Brinkman 2014; Cordner 2017; Charman 2014; Murphy and Greenhalgh 2018). We have however argued in this volume that emergency service personnel are witnessing a period of profound change due to fiscal tightening of the service budgets (NAO 2015a, b, 2017) and changing patterns of demand and workforce shortage (HMIC 2016; NHS Workforce Strategy 2017; Wankhade 2017). The impact of these changes on the mental health and well-being of the blue light workforce is becoming a cause of concern (see Wankhade 2018) as highlighted in the previous chapter. This is accompanied by new emerging terrorist threats and the ever-present risks of dealing with man-made or natural disasters. Such unprecedented changes in the societal, political and economic landscape of these services have necessitated the need to look deeper into the changing identities and dynamics of the professional cultures across these organisations. Additional pressures to increase joint working between the emergency services are also being witnessed with the police, paramedics and firefighters taking over aspects of each other's job, thus introducing a new dynamic in the inter-professional relations (Tehrani and Hesketh 2018).

Police Culture

Police services have undergone fundamental governance arrangements, some of which were highlighted earlier in the introductory chapter. Police culture(s) traditionally have been discussed in the literature, showing characteristics such as hegemony, masculinity, racism, sexism and prejudice

(Loftus 2009; Charman 2017; Reiner 2010; Martin 1980) with the "pre-occupation with crime reflected the dominance of a masculine ethos within the occupational culture" (Loftus 2010, p. 6). Occupations such as the armed forces, the police or the fire services have been characterised by a 'male bonding' culture, though such sentiments are changing with recruitment of more women (Gregory and Lees 1999). Differences of their experience between male and female police officers have also been recorded (Martin 1996). However, with the inclusion of more women, ethnic minorities, people with different sexual orientation and graduates in the police workforce, assumptions around "they are the same" are being challenged as the process of socialisation and culture does not provide a single vision of the police or a single way of doing police work (Cordner 2017, p. 11; Brown 1981; Muir 1977). Accounts of orthodox nature of police culture having considerable value also exist (Loftus 2010).

In addition to the issue of gender representation, diversity remains another big challenge for the police. A Parliamentary Report (House of Commons 2016) has revealed that the Black, Muslim and Ethnic (BME) representation in England and Wales has marginally improved from 3.6% in 2006 to 5.5% in 2015, compared to 14% of the population and 11.4% of the UK workforce respectively. The gap between police profile and the community it serves is quite big and is the worst for the Metropolitan Police Service (in London), where the BME police officer representation is 12.4% as compared to 40.2% of the population. The report has further highlighted "considerable variation across the country, with some forces having difficulty in attracting applications from BME members of the community and having a very low proportion of new BME recruits" along with poor levels of BME representation at senior levels (ibid., p. 6).

International evidence also points to the need for a more nuanced understanding of the phenomenon of police culture. Paoline and Terrill (2014) discuss the incongruent conceptualisations of police culture and also discuss features of the profession where there is an agreement among staff as well those elements which produce cultural heterogeneity. A recent study (Cordner 2017) involving survey data from about 90 US police and sheriff department highlighted a more positive aspect of staff perspectives. The study also concluded that "police culture is not necessarily problematic, it is not monolithic, and it is substantially organizational, not simply occupational or a matter of individual personal differences" (ibid., p. 21). In another study involving the German Police,

Jacobs and Keegan (2018) interviewed participants for their reactions to several large-scale planned change projects which highlighted 'ambivalence' rather than resistance due to ethical considerations of how the change affected other employees and the organisation. In a study with Australian Police officers by Brough et al. (2016), five dominant organisational culture characteristics were revealed: the police family, control, us versus them, masculinity and subcultural differences. The study also identified two dominant themes, namely reduction in social rituals and increased scrutiny.

Some accounts of police culture attribute behaviour to its unique occupational environment instead of an acknowledgement of the many similarities and interesting parallels with other professions (Foster 2005). For instance, Winslow (1998) highlighted the issues of solidarity and hierarchy in the context of the military cultures. More recently, Wankhade (2012, 2017, 2018) has drawn attention to the link between performance regime and response time targets of ambulance staff, which has implications for other public sector environments as well. Charman (2014, p. 115) raised the notion of 'cultural interoperability' between the frontline police and ambulance staff in her study, arguing that "the two occupational groups display tangible behaviours and organisational characteristics that exhibit mutual engagement, joint enterprise and communal resources".

Notwithstanding the shift seen by many commentators on the nature of such change, it is a mixed picture. A deep-rooted occupational culture resulting in suspicion and cynicism by the rank and file officers has been attributed for failure of reform process (Barton 2003). Loftus (2008, p. 756) argued that "the narratives of demise and discontent put forward by the adherents of the former operate to subordinate the spaces of representation for emerging identities and sustain an increasingly endangered culture". Charman (2017) however offers a contrasting perspective and questions whether the criticism levelled against the police culture is justified. In a longitudinal study of new police recruits who referred to themselves as a 'new breed' of police officer, she noted new and enduring cultural characteristics at the entry level of staff and concluded that a positive change is taking place in policing cultures, though such change is likely to be slow and incremental. We would like to argue that any orthodox concepts of police culture(s) would not be relevant in modern policing.

Ambulance Culture(s)

Ambulance service culture like the police services has also been historically characterised as male-dominated, hierarchical, with the tendency to blame (Commission for Health Improvement 2002) and being risk-averse (NHS Modernisation Agency 2004). Similar accounts exist in international domains (see Metz 1981; Palmer 1983) which depict the 'macho' culture in the service and levels of intensity of ambulance work.

Historically, the ambulance work has focused on the 'speed of response' to get to the patient in the quickest way through a series of response time targets, and the training needs were organised around dealing with life-threatening emergencies or major trauma (Cooper 2005; Mannon 1992; Alexander and Klein 2001; Lendrum et al. 2000; Pell et al. 2001). However with the development of more clinical skills and treatment options available to ambulance crews, there was a need to develop a new thinking (Brown et al. 2000). The first national review in the UK, *Taking Healthcare to the Patient*, set out a blueprint for the cultural transformation "from a service focusing primarily on resuscitation, trauma and acute care towards becoming the mobile health resource for the whole NHS" (Department of Health 2005, p. 5). Consequently ambulance services started to focus on broader assessment and triage tools, with greater attention to clinical education and training and through improvements in leadership (Association of Ambulance Chief executives AACE 2011; Lovegrove and Davis 2013). Notwithstanding the progress made by ambulance service to culturally transform themselves from a health arm of the emergency services to the emergency arm of the health services, it is still perceived by some as a 'call-handling' and patient transportation service (Wankhade et al. 2015; National Audit Office NAO 2011). McCann et al. (2013) highlight several challenges in transforming a 'blue collar trade' to a professional healthcare organisation.

Wankhade (2012) has documented three distinct sub-cultures in the English ambulance services, applying Schein (1996) typography. These included:

1. The culture of the 'operators'—the frontline paramedics who respond to the emergency 999 calls,
2. The culture of the 'engineers'—the emergency operations centre staff who receive the 999 calls and dispatch an appropriate response to each call and

3. The culture of the 'executives'—senior executives and the chief executives along with the executive team.

A variety of assumptions and beliefs of each of the three occupational groups reflected a multilayered nature of the organisational culture in the study, each with a particular attitude towards performance and targets. Further evidence will help to gain more insights into this phenomenon.

A need for a cultural shift has been highlighted in several policy documents and academic papers (Wankhade and Mackway-Jones 2015; McCann et al. 2013, 2015; Wankhade et al. 2015; Wankhade 2011b; DH 2005). Evidence about the impact of culture change within the ambulance services is however mixed with few empirical studies exploring specific episodes of cultural change. Without a clear set of change objectives, any culture change initiatives can lead to dysfunctional behaviours and unintended consequences. In an ethnographic study in an NHS ambulance trust, Wankhade and Brinkman (2014) highlighted six such unintended consequences of a culture change programme (see also Harris and Ogbonna 2002). A more detailed account is beyond the scope of this discussion, but involved the following:

1. *Hijacked process* of culture change due to the vagaries of operational exigencies,
2. *Cultural erosion* in which espoused ideals are subsequently eroded,
3. *Ivory tower culture change* without taking the organisational reality into account,
4. *Inattention to symbolism* by ignoring cultural sensitivities of different groups and measure of success,
5. *Ritualisation of culture* change to a periodic ritual and
6. *Behavioural compliance*—going through emotions without any major shift in deep-rooted assumption.

Thus, any meaningful analysis of a planned change initiative should take into account negative consequences while considering any positive impact of such changes. The importance of timing is also important in studying organisational culture, which is a dynamic process showing different effects over time (Harris and Ogbonna 1998; Jackson 1997). There is a clear research gap in this area, and conducting further empirical studies in different ambulance setting will be a fruitful research endeavour.

Additionally, a 'target culture' has been identified by several commentators as a significant constituent of the ambulance culture, impacting workforce behaviour and well-being (Granter et al. 2019; Wankhade et al. 2019; Heath et al. 2018; Wankhade 2011a; Radcliffe and Heath 2009; Heath and Radcliffe 2007; Bevan and Hamblin 2009). 'What's measured is what matters' (Bevan and Hood 2006) rings true in case of the ambulance services. Ambulance services have made massive progress in their path towards professionalisation. From 2020, entry to paramedic training will be through a university graduate course, and the paramedics have been given new powers to prescribe medication to reduce admissions to A&E, thus showing greater confidence in their ability to make clinical decisions. Several specialised roles for paramedics have been approved and are currently practiced across different ambulance trusts in the UK (Wankhade 2016).

Despite these advances, recent evidence (NAO 2017, p. 8; House of Commons 2017) suggests a "general consensus that commissioners, regulators and providers still place too much focus on meeting response times … and important factors other than response times require attention when managing ambulance service performance". Wankhade et al. (2018) have recently explored the phenomenon of 'cultural perpetuation', defined as "continuation of core cultural values, beliefs, and assumptions such that they become enduring in such a way that new generation of organisational members are conditioned to adopt them in responding to various organisational contingencies" (Ogbonna and Harris 2014, p. 668). The authors (Wankhade et al. 2018, p. 940) concluded from their study in a large NHS ambulance trusts that "efforts to bring about organisational change through training and culture change were countered by the continuing emphasis on response times, which perpetuated the existing, dominant culture".

Ambulance services are at a crucial point in their professionalisation journey, but the scale and pace of these changes coupled with increasing uncertainties, not least what Brexit might bring, together with increasing 999 demands, is also impacting the health and well-being of ambulance staff. A sustainable strategy will need to be devised to bring a real transformational change to their clinical practice and future direction of travel.

Fire Culture

Compared to the police and ambulance service, literature on fire service culture is rather sparse, with very few empirical studies on the organisational culture (Thomas 2015). However few accounts exist in which the fire service culture is characterised as strong and rank-based, with a highly prescriptive discipline code placing high regard to group and group membership and dominated by men (Archer 1999; Yarnal et al. 2004). Andrews and Ashworth (2018, p. 146) highlight the elements which combine to shape this culture: "high levels of workplace trade unionism (more than 80%); the close-knit and 'dirty' nature of the occupation; and homogeneity within the work force".

The notion of 'hegemonic' masculinity has been highlighted in the literature (Ward and Winstanley 2006; see Tracy and Scott 2006), and fighting fire has also been compared to 'going to war' (Greenberg 1998). The role of a modern firefighter has however evolved to include more mundane things such as dealing with floods, road traffic collisions (RTCs) and medical emergencies rather than simply putting out fires. Firefighters value teamwork, and in the team, 'boisterous male sociability' is often valued and individual performance is closely scrutinised (Hall et al. 2007). Every firefighter also joins a *Watch culture*—referred to by colours—which comprises people working in the same shift. The group bonding is reinforced by staying together in dormitories at fire stations for resting during night shifts. Not surprisingly, "firefighting is still considered to be a masculine form of heroism" (Yarnal et al. 2004, p. 686).

Similarly Andrews and Ashworth (2015) concluded in their study of the UK civil service departments that "representative organizations are especially well placed to remove barriers to organizational participation and inclusion" (p. 287). Statistics reveal that in the UK, women constituted 3.1% of total operational firefighters in 2007, and their representation at the senior positions was around 1% (Department for Communities and Local Government DCLG 2008). More recent figures show some improvement, but women still represented only 14.4% of the workforce in 2014 as compared to 12.7% of the workforce of fire brigades between 2001 and 2006 Additionally diversity remains a challenge (Bain Report 2002). During 2001–2006, ethnic minority representation was an average of 1.3% of each fire brigade workforce, which rose to 4.4% in 2014 (Andrews and Ashworth 2018, p. 149). A recent story (Guardian 2017) reported that the service is '96% white and 95% male'.

Diversity is still a challenge in a workplace dominated by men. Ward and Winstanley (2006, p. 193) highlighted the complexities and dynamics of sexual minorities living and working in the *Watch culture* in the fire service. However with the entry of more lesbian, gay, bisexual people into the service, greater openness has been seen along with support from the Fire Brigade Union (FBU) and in their policy document '*All Different, All Equal*' (FBU 2007). Similarly, with the entry of more women in firefighter roles, perceptions in a male-dominated occupational culture are gradually changing, but few tensions still remain: "The debate about women's ability to carry out firefighting's physical work is undercut with tensions around their potentially disruptive effect upon the complex hierarchies of male sociality that structure watch life" (Hall et al. 2007, p. 541).

These issues require examination within the context of some of the recent changes in the political and operational landscape of the fire service in the UK. Over the past decades, overall attendance at incidents is down by 40%, attendance at fires is down by 48%, building fires are down by 39% and the fire services 999 attendance is on a long-term low (Knight Report 2013). Such significant changes in the operational delivery of the fire service are accompanied by massive reduction of the budgets. The National Audit Office (2015b) has highlighted risks to the financial and service sustainability of the service. For instance between 2010 and 2015, funding for standalone fire and rescue authorities fell significantly on average by 28% in real terms (NAO 2015b, p. 6). While the sector had to absorb the funding reductions, savings have come largely by reducing the number of firefighters while appliances and fire stations have been largely protected (NAO 2015b). The service has done a commendable job in the prevention work which is "bolstered and enabled by its reputation as a trusted organisation" (Mansfield 2015, p. 8).

Another main feature of the fire service culture has been the dominant role, position and influence of the Fire Brigade Union (FBU). Several commentators have described the culture of workplace trade unionism (Winchester 1983). As Fitzgerald and Stirling (1999, p. 54) pointed out:

> This culture of workplace trade unionism is the key to understanding how management attempts to implement budget reductions or performance indicators are mediated and controlled. The power of the FBU does not simply reside in its numerical and organisational strength, nor does the membership rely on the leadership of particular activists. ... The trade unionism of the FBU is rooted in the 'watch' which focuses on the necessity for loyalty and team working which is translated into a consciousness of the importance of the collective.

This corporate harmony approach (Fitzgerald and Stirling 1999; see also Fitzgerald 2005) has held sway though criticism of a 'militant' approach (see Kelly 1996), also discussed in the literature.

Fire service is witnessing a 'changeable moment' in a post-Grenfell tragedy and greater public scrutiny of their role in dealing with this tragic incident. The political reforms leading to the Policing and Crime Act 2017 has enabled the Police and Crime Commissioners to make a business case for running the police and fire services together. This is already happening in about a dozen cases in England now. The budgetary cuts (and pressure from FBU) will require discussion on new service models, ranging from mergers of individual fire services (Knight 2013; Bain 2002) to the wider integration of all the emergency services or sharing excess capacity and resources with the ambulance and police service. One suggestion has been about offering solutions to cash-strapped local authorities aware of the need for earlier intervention, since "community intervention models show evidence that the (fire) service could be a key part of solving these problems" (Mansfield 2015, p. 12).

Conclusion

As our literature review suggests, organisational culture is a complex and amorphous concept which is subject to various variables/contingencies and is dynamic in nature (Wankhade et al. 2015). While this makes a systematic understanding of culture and culture change so much more difficult, it also provides exciting opportunities to delve further into this fascinating topic. Given the policy impetus to bring about a positive change in the culture of these important public service organisations, continued academic and scholarly interest will further help to unearth new dynamics.

Emergency services in many parts of the world are confronted with similar challenges of dealing with reduced budgets, greater drive towards professionalisation and similar cultural obstacles and evolving into professional organisations (Wankhade et al. 2019). Our analysis suggests that while a tangible change is happening across the blue light services, the pace of such cultural change is still slow and is sometimes thwarted by operational, political or organisational exigencies. However, with new entry routes for new recruits along with a younger, diverse workforce, professional cultures across the three services require to be understood with a new lens and open minds.

The next final chapter summarises the key findings from the themes covered in this volume and also suggests further research avenues.

REFERENCES

Ackroyd, S., & Crowdy, P. (1990). Can culture be managed? Working with 'raw' material: The case of the English slaughtermen. *Personnel Review, 19*(5), 3–13.

Alexander, D. A., & Klein, S. (2001). Ambulance personnel and critical incidents. *British Journal of Psychiatry, 178*, 76–81.

Andrews, R., & Ashworth, R. (2015). Representation and inclusion in public organizations: Evidence from the U.K. Civil Service. *Public Administration, 75*(2), 279–288.

Andrews, R., & Ashworth, R. (2018). Feeling the heat? Management reform and workforce diversity in the English fire service. In P. Murphy & G. Greenhalgh (Eds.), *Fire and rescue services: Leadership and management perspectives* (pp. 145–158). Cham: Springer.

Archer, D. (1999). Exploring "bullying" culture in the para-military organisation. *International Journal of Manpower, 20*(1/2), 94–105.

Association of Ambulance Chief Executives. (2011, June). *Taking healthcare to the patient 2: A review of 6 years' progress and recommendations for the future.* London: AACE.

Bain, G. (2002). *Independent review of the fire service.* London: Home Office.

Barton, H. (2003). Understanding occupational (sub) culture – A precursor for reform: The case of the police service in England and Wales. *International Journal of Public Sector Management, 16*(5), 346–358.

BBC. (2012, July 12). Surrey police shelves plan to privatise roles with G4S. *BBC News.* Retrieved January 10, 2019, from https://www.bbc.co.uk/news/uk-england-surrey-18813591.

Bevan, G., & Hamblin, R. (2009). Hitting and missing targets by ambulance services for emergency calls: Effects of different systems of performance measurement within the UK. *Journal of the Royal Statistical Society, 172*(Part 1), 161–190.

Bevan, G., & Hood, C. (2006). What's measured is what matters: Targets and gaming in the English public health care system. *Public Administration, 84*(3), 517–538.

Bititci, U. S., Mendibil, K., Nudurupati, S., Garengo, P., & Turner, T. (2007). Dynamics of performance measurement and organisational culture. *International Journal of Operations & Production Management, 26*(12), 1325–1350.

Brough, P., Chataway, S., & Biggs, A. (2016). You don't want people knowing you're a copper! A contemporary assessment of police organisational culture. *International Journal of Police Science & Management, 18*(1), 28–36.

Browaeys, M.-J., & Price, R. (2008). *Understanding cross-cultural management.* Harlow: FT/Prentice Hall.

Brown, M. K. (1981). *Working the street: Police discretion and the dilemmas of reform.* New York: Russell Sage Foundation.

Brown, L., Whitney, C., Hunt, R., Addario, M., & Hogue, T. (2000). Do warning lights and sirens reduce ambulance response times? *Prehospital Emergency Care, 4*(1), 70–74.

Carlström, E., & Lars-Eric, O. (2014). The association between subcultures and resistance to change – In a Swedish hospital clinic. *Journal of Health Organization and Management, 28*(4), 458–476.

Charman, S. (2013). Sharing a laugh: The role of humour in relationships between police officers and ambulance staff. *International Journal of Sociology and Social Policy, 33*(3-4), 152–166.

Charman, S. (2014). Blue light communities: Cultural interoperability and shared learning between ambulance staff and police officers in emergency response. *Policing and Society, 24*(1), 102–119.

Charman, S. (2017). *Police socialisation, identity and culture: Becoming blue.* London: Palgrave.

College of Policing. (2015). *Leadership review: Recommendations for delivering leadership at all levels.* Coventry: College of Policing. Retrieved June 17, 2018, from https://www.college.police.uk/What-we-do/Development/Promotion/the-leadership-review/Documents/Leadership_Review_Final_June-2015.pdf.

Commission for Health Improvement. (2002). What CHI has found in ambulance trusts. Retrieved from www.healthcarecommission.org.uk/NationalFindings/National.

Cooper, S. (2005). Contemporary UK paramedical training and education. How do we train? How should we educate? *Emergency Medicine Journal, 22*(5), 375–379.

Cordner, G. (2017). Police culture: Individual and organizational differences in police officer Perspectives. *Policing: An International Journal of Police Strategies & Management, 40*(1), 11–25.

Deal, T. E., & Kennedy, A. (1982). *Corporate cultures: The rites and rituals of corporate life.* Reading, MA: Addison-Wesley.

Department for Communities and Local Government. (2008). *Fire and rescue service equality and diversity strategy 2008–2018.* London: Department for Communities and Local Government. Retrieved February 2, 2019, from https://assets.publishing.service.gov.uk/government/uploads/system/uploads/attachment_data/file/7633/equalitydiversitystrategy.pdf.

Department of Health. (2005). *Taking healthcare to the patient; Transforming NHS ambulance services.* London: Department of Health.

Diefenbach, T. (2007). The managerialistic ideology of organisational change management. *Journal of Organizational Change Management, 20*(1), 126–144.

Dingwall, R., & Strangleman, T. (2005). Organizational cultures in the public services. In E. Ferlie, L. Lynn, & C. Pollitt (Eds.), *Oxford handbook of public management*. Oxford: Oxford University Press.

Dutton, J. E., Ashford, S. J., O'Neil, R., & Lawrence, K. (2001). Moves that matter: Issue selling and organizational change. *Academy of Management Journal, 44*(4), 716–736.

Ferlie, E., & Shortell, S. (2001). Improving the quality of health care in the United Kingdom and the United States: A framework for change. *Milbank Quarterly, 79*(2), 281–316.

Fire Brigade Union. (2007). *All different all equal: A review.* Surrey: Fire Brigade Union.

Fitzgerald, I. (2005). The death of corporatism? Managing change in the fire service. *Personnel Review, 34*(6), 648–662.

Fitzgerald, T. (2015, November 2). Bus firm running non-emergency patient transport pays back £1.5 m after standards reporting gaffe. *Mirror.* Retrieved January 15, 2019, from https://www.mirror.co.uk/news/uk-news/bus-firm-running-non-emergency-6754569.

Fitzgerald, I., & Stirling, J. (1999). A slow burning flame? Organisational change and industrial relations in the fire service. *Industrial Relations Journal, 30*(1), 46–60.

Foster, J. (2005). Police cultures. In T. Newburn (Ed.), *Handbook of policing* (pp. 196–227). Devon: Willan Publishing.

Granter, E., Wankhade, P., McCann, L., Hassard, J., & Hyde, P. (2019). Multiple dimensions of work intensity: Ambulance as edgework. *Work, Employment and Society, 33*(2), 280–297.

Greenberg, A. (1998). *Cause for alarm: The volunteer fire department in the nineteenth century city.* Princeton, NJ: Princeton University Press.

Gregory, G., & Lee, S. (1999). *Policing sexual assault.* London: Routledge.

Guardian. (2017, February 7). *England's fire service criticised for 'woeful' lack of diversity.* Press Association. Retrieved December 15, 2018, from https://www.theguardian.com/uk-news/2017/feb/07/england-fire-service-lack-diversity.

Hall, A., Hockey, J., & Robinson, V. (2007). Occupational cultures and the embodiment of masculinity: Hairdressing, estate agency and firefighting. *Gender, Work and Organization, 14*(6), 534–551.

Harris, L. C., & Ogbonna, E. (1998). Employee responses to culture change efforts. *Human Resource Management Journal, 8*(2), 78–92.

Harris, L. C., & Ogbonna, E. (2002). The unintended consequences of culture interventions: A study of unexpected outcomes. *British Journal of Management, 13*(1), 31–49.

Hatch, M. J. (1993). The dynamics of organizational culture. *Academy of Management Review, 18*(4), 657–693.

Hatch, M. J. (2000). The cultural dynamics of organizing change. In N. M. Ashanasy, C. P. M. Wilderom, & M. F. Peterson (Eds.), *Handbook of organizational culture and climate* (pp. 245–260). Thousand Oaks, CA: Sage.

Heath, G., & Radcliffe, J. (2007). Performance measurement and the English ambulance service. *Public Money & Management, 27*(3), 223–228.

Heath, G., Radcliffe, J., & Wankhade, P. (2018). Performance management in the public sector: The case of the English ambulance service. In E. Harris (Ed.), *The Routledge companion to performance management and control* (pp. 417–438). London: Routledge.

HMIC. (2016). *The state of policing. The annual assessment of policing in England and Wales 2015*. London: Her Majesty's Inspectorate of Policing. Retrieved January 10, 2019, from www.justiceinspectorates.gov.uk/hmicfrs/publications/state-of-policing-the-annual-assessment-of-policing-in-england-and-wales-2016.

House of Commons. (2016). *Police diversity*. First Report of Session 2016–17, HC 27. House of Commons Home Affairs Committee. London: Stationery Office.

House of Commons Committee of Public Accounts. (2017). *NHS ambulance services*. Sixty-second Report of Session 2016–17, HC 1035, April 2017.

Jackson, S. (1997). Does organizational culture affect out-patient DNA rates? *Health Manpower Management, 23*(6), 233–236.

Jacobs, G., & Keegan, A. (2018). Ethical considerations and change recipients' reactions: 'It's not all about me. *Journal of Business Ethics, 152*, 73–90.

Jorritsma, P. Y., & Wilderom, C. (2012). Failed culture change aimed at more service provision: A test of three agentic factors. *Journal of Organizational Change Management, 25*(3), 364–391.

Kelly, J. (1996). Union militancy and social partnership. In P. Ackers, C. Smith, & P. Smith (Eds.), *The new workplace and trade unionism*. London: Routledge.

Knight, K. (2013). *Facing the future: Findings from the review of efficiencies and operations in fire and rescue authorities in England*. London: Department for Communities and Local Government.

Konteh, F. H., Mannion, R., & Davies, H. T. O. (2010). Understanding culture and culture management in the English NHS: A comparison of professional and patient perspectives. *Journal of Evaluation in Clinical Practice, 17*, 111–117.

Legge, K. (1994). Managing culture: Fact or fiction. In K. Sisson (Ed.), *Personnel management: A comprehensive guide to theory and practice in Britain* (pp. 397–433). Oxford: Blackwell.

Lendrum, K., Wilson, S., & Cooke, M. W. (2000). Does the training of ambulance personnel match the workload seen? *Pre-hospital Immediate Care, 4*(1), 7–10.

Lister, S. (2014). Scrutinising the role of the police and crime panel in the new era of police governance in England and Wales. *Safer Communities, 15*(1), 22–31.

Loftus, B. (2008). Dominant culture interrupted: Recognition, resentment and the politics of change in an English police force. *The British Journal of Criminology, 48*(6), 756–777.

Loftus, B. (2009). *Police culture in a changing world.* Oxford: Oxford University Press.

Loftus, B. (2010). Police occupational culture: classic themes, altered times. *Policing and Society: An International Journal of Research and Policy, 20*(1), 1–20.

Lok, P., Westwood, R., & Crawford, J. (2005). Perceptions of organisational sub-culture and their significance for organisational commitment. *Applied Psychology, 54*(4), 490–514.

Lovegrove, M., & Davis, J. (2013). *Maximising paramedics' contribution to the delivery of high quality and cost effective patient care.* High Wycombe: Buckinghamshire New University.

Mannion, R., Davies, H., & Marshall, M. (2005). *Cultures for performance in health care.* Maidenhead: Open University Press.

Mannon, J. M. (1992). *Emergency encounters: EMTs and their work.* Boston, MA: Jones and Bartlett.

Mansfield, C. (2015). *Fire works: A collaborative way forward for the fire and rescue service.* London: New Local Government Network (NLGN).

Marsh, S. (2018, September 20). Huge rise in ambulance callouts to deal with spice users. *The Guardian.* Retrieved January 14, 2019, from https://www.theguardian.com/politics/2018/sep/20/huge-rise-in-ambulance-callouts-to-deal-with-spice-users.

Martin, S. E. (1980). *Breaking and entering: Policewomen on patrol.* Berkeley, CA: University of California Press.

Martin, J. (1992). *Culture in organizations: Three perspectives.* New York: Oxford University Press.

Martin, C. (1996). The impact of equal opportunities policies on the day-to-day experiences of women police constables. *British Journal of Criminology, 36*(4), 510–528.

McCann, L., Granter, E., Hyde, P., & Hassard, J. (2013). Still blue-collar after all these years? An ethnography of the professionalization of emergency ambulance work. *Journal of Management Studies, 50*(5), 750–776.

McCann, L., Hassard, J., Granter, E., & Hyde, P. (2015). 'You can't do both: something will give': Limitations of the targets culture in managing UK health care workforces. *Human Resource Management, 54*(5), 773–791.

Meek, V. L. (1988). Organizational culture: Origins and weaknesses. *Organization Studies, 9*(4), 453–473.

Metz, D. L. (1981). *Running hot: Structure and stress in ambulance work.* Cambridge, MA: ABT Books.

Morgan, G. (1986). *Images of organisation.* London: Sage.

Muir, W. K., Jr. (1977). *Police: Streetcorner politicians.* Chicago, IL: University of Chicago Press.

Murphy, P., & Greenhalgh, G. (Eds.). (2018). *Fire and rescue services: Leadership and management perspectives.* Cham: Springer.

National Audit Office. (2015a). *Financial sustainability of police forces in England and Wales.* Retrieved February 18, 2018, from www.nao.org.uk/wp-content/uploads/2015/06/Financial-sustainability-of-policeforces.pdf.

National Audit Office. (2015b). *Impact of funding reductions on fire and rescue services.* Retrieved February 18, 2018, from www.nao.org.uk/report/impact-of-funding-reductions-on-fire-and-rescue-services/.

National Audit Office NAO. (2011). *Transforming NHS ambulance services.* London: Stationery Office.

National Audit Office NAO. (2017). *NHS ambulance services.* HC 972, Session 2016–17. London: Stationery Office.

NHS & Health Education England. (2017). *Facing the facts, shaping the future: A draft health and care workforce strategy for England to 2027.* NHS England: Leeds. Retrieved December 15, 2018, from https://www.hee.nhs.uk/print-pdf/our-work/planning-commissioning/workforce-strategy.

NHS Modernisation Agency. (2004). *Driving change: Good practice guidelines for PCTs on commissioning arrangements for emergency ambulance services and non-emergency patient services.* Retrieved from www.dh.gov.uk/prod_consum_dh/groups/dh_digitalassets/@dh/@en/documents/digitalasset/dh_4112351.pdf.

Ogbonna, E., & Harris, L. C. (2002). Managing organisational culture: Insights from the hospitality industry. *Human Resource Management Journal, 12*(1), 33–53.

Ogbonna, E., & Harris, L. C. (2014). Organizational cultural perpetuation: A case study of an English premier league football club. *British Journal of Management, 25*(4), 667–686.

Palmer, C. E. (1983). 'Trauma junkies' and street work: Occupational behavior of paramedics and emergency medical technicians. *Journal of Contemporary Ethnography, 12*(2), 162–183.

Paoline, E. A., & Terrill, W. (2014). *Police culture: Adapting to the strains of the job.* Durham, NC: Carolina Academic Press.

Pell, J. P., Sirel, J. M., Marsden, A. K., Ford, I., & Stuart, M. C. (2001). Effect of reducing ambulance response times on deaths from out of hospital cardiac arrest: Cohort study. *British Medical Journal, 322*(7299), 1385–1388.

Peters, T., & Waterman, R. (1982). *In search of excellence.* New York: Random House.

Radcliffe, J., & Heath, G. (2009). Ambulance calls and cancellations: Policy and implementation issues. *International Journal of Public Sector Management*, 22(5), 410–422.

Reiner, R. (2010). *The politics of the police* (4th ed.). Oxford: Oxford University Press.

Sackmann, S. A. (1992). Culture and subculture: An analysis of organisational knowledge. *Administrative Science Quarterly, 37*(1), 140–161.

Schein, E. H. (1985). *Organisational culture and leadership.* San Francisco, CA: Jossey-Bass.

Schein, E. H. (1996). Three cultures of management: The key to organizational learning. *Sloan Management Review, 38*(1), 9–20.

Schein, E. H. (2004). *Organizational culture and leadership.* San Francisco, CA: Jossey-Bass.

Smircich, L. (1983). Concepts of culture and organizational analysis. *Administrative Science Quarterly, 28*(3), 339–358.

Tehrani, N., & Hesketh, I. (2018). The role of psychological screening for emergency service responders. *International Journal of Emergency Services.* https://doi.org/10.1108/IJES-04-2018-0021.

Thomas, A. (2015, February). *Independent review of conditions of service for fire and rescue staff in England.* London: The Stationery Office. Retrieved from http://www.fitting-in.com/reports/Thomas_Review_independent%20review%20of%20fire%20service.pdf.

Tracy, S. J., & Scott, C. (2006). Sexuality, masculinity, and taint management among firefighters and correctional officers: Getting down and dirty with "America's Heroes" and the "Scum of Law Enforcement". *Management Communication Quarterly, 20*(1), 6–38.

Waddington, P. A. J. (1999). Police (canteen) sub-culture: An appreciation. *British Journal of Criminology, 39*(2), 287–309.

Wankhade, P. (2011a). Performance measurement and the UK emergency ambulance service: Unintended Consequences of the ambulance response time targets. *International Journal of Public Sector Management, 24*(5), 384–402.

Wankhade, P. (2011b). Emergency services in austerity: Challenges, opportunities and future perspectives for the ambulance service in the UK. *Ambulance Today, 8*(5), 13–15.

Wankhade, P. (2012). Different cultures of management and their relationships with organizational performance: Evidence from the UK ambulance service. *Public Money & Management, 32*(5), 381–388.

Wankhade, P. (2016). Staff perceptions and changing role of pre-hospital profession in the UK ambulance services. *International Journal of Emergency Services, 5*(2), 126–144.

Wankhade, P. (2017, July 13). How to reboot Britain's fractured emergency services. *The Conversation.* Retrieved from https://theconversation.com/how-to-reboot-britains-fractured-emergency-services-79528.

Wankhade, P. (2018). The crisis in NHS ambulance services in the UK: Let's deal with the 'elephants in the room'! *Ambulance Today, 15*(1), 13–17.

Wankhade, P., & Brinkman, J. (2014). The negative consequences of culture change management: Evidence from a UK NHS ambulance service. *International Journal of Public Sector Management, 27*(1), 2–25.

Wankhade, P., & Mackway-Jones, K. (Eds.). (2015). *Ambulance services: Leadership and management perspectives.* New York: Springer.

Wankhade, P., Radcliffe, J., & Heath, G. (2015). Organisational and professional cultures: An ambulance perspective. In P. Wankhade & K. Mackway-Jones (Eds.), *Ambulance services: Leadership and management perspectives* (pp. 65–80). New York: Springer.

Wankhade, P., Heath, G., & Radcliffe, J. (2018). Culture change and perpetuation in organisations: Evidence from an English ambulance service. *Public Management Review, 20*(6), 934–948.

Wankhade, P., McCann, L., & Murphy, P. (2019). *Critical perspectives on the management and organization of emergency services.* New York: Routledge.

Ward, J., & Winstanley, D. (2006). Watching the watch: The UK fire service and its impact on sexual minorities in the workplace. *Gender, Work and Organization, 13*(92), 193–219.

Winchester, D. (1983). Industrial relations in the public sector. In G. Bain (Ed.), *Industrial relations in Britain.* London: Blackwell.

Winslow, D. (1998). Misplaced loyalties: The role of military culture in the breakdown of discipline in peace operations. *Canadian Review of Sociology and Anthropology, 35*(3), 345–367.

Winsor, T. P. (2012). *Independent review of police officer and staff remuneration and conditions final report* (Vol. 1, Cm 8325-I). London: Stationery Office.

Woollard, M. (2006). The role of the paramedic practitioner in the UK. *Australian Journal of Paramedicine, 4*(1), 11.

Woollard, M. (2009). Professionalism in UK paramedic practice. *Australian Journal of Paramedicine, 7*(4), 9.

Yarnal, C. M., Dowler, L., & Hutchinson, S. (2004). Don't let the bastards see you sweat: Masculinity, public and private space, and the volunteer firehouse. *Environment and Planning A: Economy and Space, 36*(4), 685–699.

Conclusion: Collaboration and Governance: It's Very Much About 'Process' and 'People'!

Abstract This final chapter draws together the key findings from the previous chapters and summarises the core themes covered in this Palgrave Pivot volume on *Collaboration and Governance in the Emergency Services: Issues, Opportunities and Challenges*. Issues around improving and maintaining 'trust' and finding a more suitable and distributed leadership style pose a significant challenge for the emergency services. Similarly, the rise in mental health, stress and post-traumatic stress disorder (PTSD) cases in emergency workers has serious implications for organisational productivity and resilience. With the recruitment of a younger and diverse workforce, some of the traditional and somewhat negative connotations of organisational and occupational cultures are also changing. Therefore, we conclude that both 'process' and 'people' are central to any reform, and the success of the collaboration and governance agenda will inevitably depend on how the collaboration process is led and managed, taking into account the interests of people who would make the partnerships work. Implications for future research are also highlighted.

Keywords Emergency services • Process • People • Collaboration • Governance • Research • Complexity • Culture

© The Author(s) 2020   127
P. Wankhade, S. Patnaik, *Collaboration and Governance in the Emergency Services*,
https://doi.org/10.1007/978-3-030-21329-9_7

Scope and Background

Emergency services are witnessing massive structural, organisational and financial changes while adopting greater business practices and models (see NAO 2015, 2017; Granter et al. 2015) and are witnessing a period of unprecedented global changes within the backdrop of increased terrorist threat, new forms of crime, reduced workforce, cut in organisational budgets and a 24/7 media and external scrutiny (Tehrani and Hesketh 2018; Wankhade 2017). However, the enduring public image of a heroic, friendly and important public service will continue to grow in a climate of heightened risk and new threats (McCann et al. 2019).

The various themes covered in the chapters in this book are underpinned by academic understanding and empirical findings. They also illustrate both pockets of good practice and significant structural, organisational and policy changes impacting the collaboration and governance agenda in the emergency services. We have noted the positive developments, especially around the legislative and policy framework (such as The Policing and Crime Act 2017, the role of the Police and Crime Commissioner [PCC], JESIP [Joint Emergency Services Interoperability Programme] principles), to drive collaboration. Equally, we have articulated our concerns on the lack of progress made in tackling workforce health and well-being issues and the negative consequences of operational pressures (Sky News 2018; Wankhade et al. 2018).

In Chap. 2, we discussed the conceptual understanding of collaborations and its implications for the emergency services. In the broad domain of public sector management, inter-organisational collaborations are acknowledged as the most appropriate strategy to address some of the 'wicked' societal challenges (Crosby and Bryson 2018; Vangen and Huxham 2012). In this respect, different scholars have identified the antecedent conditions (Bryson et al. 2006; Thomson and Perry 2006; Ansell and Gash 2008), the governance mechanisms (Provan and Kenis 2008; Emerson et al. 2012; Gollagher and Hartz-Karp 2013) and leadership and management challenges (Connelly 2007; Silvia and McGuire 2010; Cristofoli et al. 2017). However we observe three distinct limitations in this body of research. First, notwithstanding greater attention paid to inter-organisational collaborations over the last two decades (Bryson et al. 2015), there is significant gap in research on collaboration between the emergency service blue light organisations. Interestingly, the collaboration between blue light organisations is not a new phenomenon, either in the UK or in other countries in the Northern Hemisphere. Individually, the police force, fire and rescue services and ambulance services have been collaborating for a long time, particularly in responding to emergency situations.

However, the drive for collaborations in the current institutional context is vastly different. The Knight Report (2013) made a case for greater collaborations amongst the blue light organisations, as an approach to deliver services in a more efficient and cost-effective manner. Put simply, the collaboration is not in response to a specific 'emergency situation', rather in a broader, day-to-day basis. The Policing and Crime Act 2017 made it mandatory for the Police and Crime Commissioners to actively pursue collaboration between the three services. Therefore, in this context, the antecedent conditions as well as the mechanisms adopted to form collaborations amongst the blue light organisations demand further exploration. Second, the nature and orientation of the collaborations amongst the blue light organisations, we observe, is significantly different as compared to other organisations. The Policing and Crime Act 2017 enables the PCCs to take on the functions and duties of Fire and Rescue Authorities and create a single employer for police and fire personnel. In this respect, the police forces and the fire and rescue services will exist under one employer. So, in essence, the collaboration, in question, is between two different entities within the same organisation and therefore further studies on emanating structural (governance) and relational (culture, trust, interpersonal relationship) are needed.

Recent insights point to some of the critical challenges that such collaborations will face as the phenomenon gets institutionalised (see Charman 2015). The third limitation of research on inter-organisational collaboration is more of thematic nature. A significant body of research has paid attention to different aspects and dimensions. However, there is a distinct gap in studies that have longitudinally explored the phenomenon (Bryson et al. 2015; Vangen et al. 2015). Adopting a process perspective, we argue is central to developing better insights and understanding on various issues, particularly how the collaborations are led and managed and how they adapt to different structural and relational internal and external contingencies over time.

In Chap. 3, we argued that deriving value from inter-organisational strategic partnerships remains one of the most critical challenges for the collaborating organisations. Therefore, it is considered that organisations that possess collaborative capabilities derive most value from their strategic collaborations. Collaborative capabilities are organisational capabilities and they pertain to the capacity of organisations to form, maintain and manage their respective inter-organisational partnerships (Kale and Singh 2009; Kohtamaki et al. 2018). Capacities related to coordination tasks and activities, communication between members of organisations belonging to different groups and teams and building and sustaining interpersonal relationships are identified as the key components of collaborative capabilities. Scholars also concur that collaborative capabilities develop from prior experience, which also highlight the significance of learning mechanisms that an organisation adopts to internalise learning from its various collaborations (Powell 1998; Wang and Rajagopalan 2015); in this respect many scholars allude to the influential role a dedicated team of managers who facilitate internalisation of learning (Kale and Singh 2009; Gulati et al. 2012). These issues have not been adequately explored in the context of collaborations in the public sector organisations in general and certainly in the case of emergency or blue light organisations in particular.

Considering the complexities associated with inter-organisational collaborations, particularly involving public sector organisations, there has been growing interest to explore relevance of existing leadership theories for these arrangements. The leadership approach has evolved in the case of blue light organisations, from conventional trait-/leader-centric approach of leadership to transformational leadership to more recently shared leadership, which emphasise involvement of organisational members in the leadership process. In essence, the focus has been on understanding what type of leadership approach fits the organisations in the current time (Crosby and Bryson 2018). However, extant studies show that in some instances, particularly in case of police forces, organisational members prefer some aspects of trait and behaviour leadership (Davis and Bailey 2018). However, there is a distinct view that in case of collaborative relationships, a different leadership approach, namely the integrative leadership approach (Page 2010; Malin and Hackmann 2019), is more suitable. The essence of integrative leadership pertains to the capacity of the leaders to integrate the efforts of the collaborating organisations towards achieving shared outcomes.

The collaboration between blue light organisations, taking into account the evolving organisational structures with the police forces and the fire and rescue services operating under one single employer model, presents a different dilemma to the leaders. Particularly it raises the question regarding who should assume the leadership of integrating the organisations?

Although the onus of the integrating should rest with the Police and Crime Commissioners (PCCs), there is lack of clarity as far as their role and the role of the Police Chief Constables are concerned. Hence, it is critical that longitudinal approach is adopted to further explore how the newly created single employer model organisations are functioning in practice and how leadership manifests in these organisations.

In Chap. 4, we highlighted the significance of 'trust'. Inter-organisational collaborations are dynamic social systems which evolve over a period of time and are underpinned by repeated interactions between key organisational members. However, such strategic partnerships are inherently unstable due to differing partner goals and objectives and perception of outcomes, cultural differences and different strategic and organisational practice including decision-making processes; as a consequence, management of inter-organisational collaborations presents critical challenges to organisational leaders. Against this backdrop, 'trust' assumes significance in creating conditions for partners to achieve their mutual objectives (Villena et al. 2019). Trust, by nature, is a multidimensional construct, in the sense that trust relationship between organisations is dependent on trust relationship between key organisational members, often called as the boundary spanners (Vanneste 2016). At the same time, trust is also multifaceted that includes goodwill trust and competence trust, with each potentially influencing the performance of the inter-organisational partnerships (Das and Teng 2000; Gulati and Sytch 2008).

In specific context to strategic partnerships between the blue light emergency organisations, both the multidimensional and multifaceted aspects of trust are relevant and need critical investigation. One could argue that JESIP interoperability principles act as building block for the development of competence trust in the sense that the principles, more or less, delineate a routine of how the organisations could function in an emergency situation. However, the collaborating organisations have to work towards developing goodwill trust, and in that respect the senior leadership will play a critical role. Charman (2015) has attempted to sensitise critically the challenges that the leaders will face in bridging the cultural barriers between the police staff and the firefighters. The centrality of interpersonal relationships in the development of trust brings forth the importance of the concept of psychological contract. As such, each of the blue light organisations has experiences of significant reduction in financial and manpower resources. Against this backdrop, Conway et al. (2014) highlight the discourse of 'doing more with less' that provides the underpinning logic for the working environment in the UK public sector, including in the blue light organisations.

The narrative of achieving 'efficiency' is the most dominant rationale for the formation of blue light collaborations, and efficiency achievement is underpinned by structural alignment of the back office activities of the three blue light organisations and the joint use of physical assets. Different research shows the detrimental impact of breach in psychological contract, which in essence refers to the trust relationship between an employee and the employers, on employee motivation and on inclination to share and co-develop knowledge (see Verburg et al. 2018). In many respect, the blue light collaborations represent a new way of working and have implications akin to organisational change and transformation. Therefore, it is absolutely critical that organisational leaders focus attention on some of the most pertinent 'people'-related issues.

In Chap. 5, we highlighted the state of affairs of emergency services workforce. On the issue of workforce health and well-being, we documented several challenges and shared our concerns about the 'stress crises' currently being witnessed across the ambulance, police and fire services. The NHS Draft Workforce Strategy published in 2017 painted a grim picture of the workforce situation in the NHS ambulance services and highlighted that of 17,000 nurses, midwives and allied health professionals (AHPs) who went on to the professional registers between 2015 and 2017 only 7000 joined the NHS and suggested that the primary reason for the staff leaving the NHS was the 'growing pressure' they were experiencing in the workplace (Public Health England 2017). Sickness absence is also the highest in the ambulance services (Wankhade 2016), and the service was reported to be far worse than other NHS organisations in terms of discrimination and equal opportunities (Vize 2018). Instances of harassment and bullying are also reported in the police services (Williams 2015; Hales et al. 2015), the ambulance services (Care Quality Commission CQC 2018) and the fire services (Drury 2016).

Some of the recent policy and academic papers further confirm the pressures of work in the emergency services workforce. For instance, the National Audit Office (2017, p. 5) highlighted a resourcing challenge for the NHS Ambulance Trusts that is limiting their ability to meet rising demand and the challenge to recruit the staff and then retain them. The report further states that "the reasons people cite for leaving are varied and include pay and reward, and the stressful nature of the job" (ibid., p. 7). In our own research, we have come to similar conclusions and found that "while the ambulance services are 'professionalizing', but as work in ambulance trusts continues to intensify, issues over dignity, staff retention and the meaning of work are becoming ever more challenging" (Wankhade et al. 2018; Granter et al. 2019, p. 280). According to research carried out by the mental health charity *MIND*, "85% of fire and rescue staff and volunteers have experienced stress and poor mental health at work, but are less likely to take time off work as a result compared to the general workforce" (MIND 2015). Similarly, according to a recently published *Police Federation Demand, Capacity and Welfare Survey 2018*, of over 18,000 police staff, almost 90% of respondents indicated that they "generally don't have enough officers to manage the demands faced by their team or unit, whilst 83.2% felt that they did not have enough officers to do their job properly" (Elliott-Davis 2019, p. 12; Hall et al. 2010).

We highlighted the changing work and demand patterns in the three emergency services in our introduction chapter. The ever-changing landscape of emergency services, coupled with emergence of new and sophisticated nature of crimes, further contributes to this complexity. While we acknowledge the work done to deal with this issue, our analysis suggests that unless workforce health and well-being is made a central plank in the ongoing professionalisation and reform agenda, meaningful results may not be achieved.

In Chap. 6, we dealt with the fascinating but controversial nature of organisational and professional culture and its role in bringing organisational change. The usefulness of understanding of organisational culture and its impact on employees in bringing a meaningful and discernible change is increasingly being discussed in the mainstream management literature (Ogbonna and Harris 2015; McCalman and Potter 2015; Jorritsma and Wilderom 2012; Harris and Ogbonna 1998). Similar accounts are now emerging in the context of the emergency services wherein the role of professional and occupational cultures is being explored, providing greater understanding of this important issue in different organisational settings (see Wankhade and Brinkman 2014; Wankhade 2012; Mansfield 2015; Charman 2014; Loftus 2010). In such works, the traditional notions of a command and control culture, accompanied by a tendency to blame, resistance to change and being risk-averse, are slowly giving way to a more nuanced understanding of this phenomenon.

Several factors have contributed to a more 'positive' aspect of occupational culture(s) with the emergency services. For instance, recruitment of a younger, more ethnically diverse workforce is challenging traditional assumptions of 'being the same' for different staff groups in the police services (Charman 2017; Cordner 2017; Paoline and Terrill 2014). Empirical studies have also documented 'ambivalence' rather than 'resistance' of police staff to large-scale planned change projects (see Jacobs and Keegan 2018). There is also a similar shift in perception of how cultural change in ambulance services can bring meaningful impact (Granter et al. 2019; Wankhade et al. 2018; Wankhade 2018) notwithstanding the history of a target culture in the organisation (Heath et al. 2018; Wankhade 2011, 2018; NAO 2011, 2017; McCann et al. 2013, 2015). Within the fire services as well, the notion of 'hegemonic' masculinity highlighted in the literature is slowly changing in a service characterised by a 'Watch' culture (Yarnal et al. 2004; The Knight Review 2013; Ward and Winstanley 2006). For instance, Mansfield (2015, p. 8) argues that the service has done a commendable job in the prevention work which is "bolstered and enabled by its reputation as a trusted organisation".

This is not to suggest that everything is fine in these organisations. Representation of ethnic minorities and recruitment of more women continue to be problems in the fire services despite the best efforts to make these organisations representative of the community they serve (see Andrews and Ashworth 2018; Guardian 2017). Recruitment of BME candidates, including their career progression prospects in the police, remains a major problem (Dodd 2019) and was a subject of a major Parliamentary Report (House of Commons 2016). Discrimination and harassment has been identified as a major issue by ambulance staff and needs immediate organisational and policy attention (Vize 2018; The King's Fund 2015).

Conclusion and Further Research Implications

Two key issues stand out from the coverage of these themes. The issue of '*process*' seems central in our analysis of maintaining and sustaining effective collaborations, both within and between the emergency services. For instance, the PCC-led 'single employer model' necessitates developing suitable 'processes' within their respective organisational systems so that senior leaders and the senior leadership (PCCs, Chief Constables, Chief Fire Officers and the Ambulance Chief Executives), take the onus for creating collaborative capabilities in their respective organisations. This also holds true for developing sustainable leadership across these organisations (Alyn, 2010). Success of collaborations amongst blue light organisations as we argued in Chap. 2 is not simply about why and under what institutional conditions they are formed; rather, it is about how each of these collaborations is managed over time in terms of the processes developed to build, nurture and support those networks.

Similarly, our analysis of the dire situation surrounding the workforce health and well-being points to the '*people*' and their importance to the success of any modernisation agenda. Several official reports (NAO 2015, 2017; HMIC 2016; House of Commons 2016; House of Commons Committee of Public Accounts 2017) have highlighted the pressure of work on emergency services staff. Notwithstanding the positive change in the organisational cultures we have alluded to in Chap. 6, especially on account of a younger, educated and a more diverse workforce, few challenges refuse to go. The issue of BME representation is one example, prompting calls from senior police leaders to allow "positive race discrimination in favour of minority ethnic recruits" (Dodd 2019a), and admissions by senior leaders that the police will be "disproportionately white for at least another 100 years at the current rate of progress" (Dodd 2019b), highlights some real 'people' issue which still need to be worked out. A clear process is needed in turn to deal with the people issues.

Several research avenues arise from our analysis of the themes covered in this book. The PCC model to jointly run the police and fire services in England is gathering steam, but as we argued in Chap. 2, the social aspects underpinning inter-organisational collaboration are intertwined with the governance aspects and more empirical research to directly study the processes of change and transformation of collaborations can unearth additional evidence on this issue. Greater collaboration, as we have argued, necessitates a different approach to leadership development which needs to be facilitated at multiple levels within the organisations. Similarly, development of collaborative culture in organisations will inevitably involve cultivating future leaders, who will encourage greater collaborative within and amongst organisations. Recent evidence (Police Federation 2019; Granter et al. 2019; Choi et al. 2016) suggests that the growing intensity of work in the emergency services continues to have a detrimental effect on the workforce, something we discussed at length in Chap. 5. More empirical evidence, in different settings, will help us to understand the issues of sickness absence, stress and post-traumatic stress disorder (PTSD) in the emergency services workforce across the three main services.

We sincerely hope that the backdrop of themes covered in this volume provides a more meaningful, subtle and honest context to our understanding of the collaboration and governance agenda in the emergency services.

But like any other research project we are also conscious that the canvass of our central theme is quite big to be captured within the scope of this short Pivot volume. We would like to invite emergency service practitioners, scholars, students and researchers to explore the fascinating context of the emergency services for their future work and for building upon the evidence base on the subject matter of this book.

REFERENCES

Alyn, K. (2010, September). Transformational leadership in the fire services: Identifying the needs, motives and values of leaders and flowers. *Fire House*. Retrieved February 10, 2019, from http://www.firehouse.com/article. 10467048/transformational-leadership-in-the-fire-service.

Andrews, R., & Ashworth, R. (2018). Feeling the heat? Management reform and workforce diversity in the English fire service. In P. Murphy & G. Greenhalgh (Eds.), *Fire and rescue services: Leadership and management perspectives* (pp. 145–158). Cham: Springer.

Ansell, C., & Gash, A. (2008). Collaborative governance in theory and practice. *Journal of Public Administration Research and Theory, 18*(4), 543–571.

Bryson, J. M., Crosby, B. C., & Stone, M. M. (2006). The design and implementation of cross-sector collaborations: Propositions from the literature. *Public Administration Review, 66*, 44–55.

Bryson, J. M., Crosby, B. C., & Stone, M. M. (2015). Designing and implementing cross sector collaborations: Needed and challenging. *Public Administration Review, 75*, 647–663.

Care Quality Commission CQC. (2018). *Inspection report–South East Coast ambulance service NHS foundation trust*. London: CQC.

Charman, S. (2014). Blue light communities: Cultural interoperability and shared learning between ambulance staff and police officers in emergency response. *Policing and Society, 24*(1), 102–119.

Charman, S. (2015). Crossing cultural boundaries: Reconsidering the cultural characteristics of police officers and ambulance staff. *International Journal of Emergency Services, 4*(2), 158–176.

Charman, S. (2017). *Police socialisation, identity and culture: Becoming blue*. London: Palgrave.

Choi, B., Schnall, P., & Dobson, M. (2016). Twenty-four-hour work shifts, increased job demands, and elevated blood pressure in professional firefighters. *International Archives of Occupational and Environmental Health, 89*(7), 1111.

Connelly, D. R. (2007). Leadership in the collaborative interorganizational domain. *International Journal of Public Administration, 30*(11), 1231–1262.

Conway, N., Kiefer, T., Hartley, J., & Briner, R. B. (2014). Change and psychological contract breach. *British Journal of Management, 25*, 737–754.

Cordner, G. (2017). Police culture: Individual and organizational differences in police officer Perspectives. *Policing: An International Journal of Police Strategies & Management, 40*(1), 11–25.

Cristofoli, D., Meneguzzo, M., & Riccucci, N. (2017). Collaborative administration: The management of successful networks. *Public Management Review, 19*(3), 275–283.

Crosby, B. C., & Bryson, J. M. (2018). Why leadership of public leadership research matters: And what to do about it. *Public Management Review, 20*(9), 1265–1286.

Das, T. K., & Teng, B.-S. (2000). Instabilities of strategic alliances: An internal tensions perspective. *Organization Science, 11*(1), 77–101.

Davis, C., & Bailey, D. (2018). Police leadership: The challenges for developing contemporary practice. *International Journal of Emergency Services, 7*(1), 13–23.

Dodd, V. (2019a, February 22). Police leader calls for laws to allow positive race discrimination. *The Guardian*. Retrieved February 22, 2019, from https://www.theguardian.com/uk-news/2019/feb/22/police-leader-calls-for-laws-to-allow-positive-race-discrimination.

Dodd, V. (2019b, February 19). Met disproportionately white for another 100 years – Police leaders. *The Guardian*. Retrieved February 19, 2019, from https://www.theguardian.com/uk-news/2019/feb/19/met-police-disproportionately-white-for-another-100-years.

Drury, I. (2016, May 25). Theresa May slams fire service chiefs for allowing 'bullying and harassment' to flourish as she unveils sweeping reforms. *Mail Online*. Retrieved November 8, 2018, from https://www.dailymail.co.uk/news/article-3609247/Theresa-slams-fire-service-chiefs-allowing-bullying-harassment-flourish-unveils-sweeping-reforms.html.

Elliott-Davis, M. (2019). PFEW demand, capacity and welfare survey 2018-Headline Statistics, December 2018. *Police Federation*. Retrieved February 20, 2019, from http://www.polfed.org/documents/DemandCapacityandWelfareSurveyHeadlineStatistics2018-06-02-19-V1.pdf.

Emerson, K., Nabatchi, T., & Balogh, S. (2012). An integrative framework for collaborative governance. *Journal of Public Administration Research and Theory, 22*, 1–29.

Gollagher, M., & Hartz-Karp, J. (2013). The role of deliberative collaborative governance in achieving sustainable cities. *Sustainability*, (6), 2343–2366.

Granter, E., McCann, L., & Boyle, M. (2015). Extreme work/normal work: Intensification, storytelling, and hypermediation in the (re)construction of 'the New Normal'. *Organization, 22*(4), 443–456.

Granter, E., Wankhade, P., McCann, L., Hassard, J., & Hyde, P. (2019). Multiple dimensions of work intensity: Ambulance as edgework. *Work, Employment and Society, 33*(2), 280–297.

Guardian. (2017, February 7). *England's fire service criticised for 'woeful' lack of diversity.* Press Association. Retrieved December 15, 2018, from https://www.theguardian.com/uk-news/2017/feb/07/england-fire-service-lack-diversity.

Gulati, R., & Sytch, M. (2008). Does familiarity breed trust? Revisiting the antecedents of trust. *Managerial and Decision Economics, 29*(2–3), 165–190.

Gulati, R., Wohlgezogen, F., & Zhelyazkov, P. (2012). The two facets of collaboration: Cooperation and coordination in strategic alliances. *Academy of Management Annals, 6,* 531–583.

Hales, G., May, T., Belur, J., & Hough, M. (2015). Chief officer misconduct in policing: An exploratory study. In *College of Policing.* Ryton-on-Dunsmore: College of Police.

Hall, G. B., Dollard, M. F., Tuckey, M. R., Winefield, A. H., & Thompson, B. M. (2010). Job demands, work–family conflict, and emotional exhaustion in police officers: A longitudinal test of competing theories. *Journal of Occupational and Organizational Psychology, 83,* 237–250.

Harris, L. C., & Ogbonna, E. (1998). Employee responses to culture change efforts. *Human Resource Management Journal, 8*(2), 78–92.

Heath, G., Radcliffe, J., & Wankhade, P. (2018). Performance management in the public sector: The case of the English ambulance service. In E. Harris (Ed.), *The Routledge companion to performance management and control* (pp. 417–438). London: Routledge.

HMIC. (2016). *The state of policing; the annual assessment of policing in England and Wales 2015.* London: Her Majesty's Inspectorate of Policing. Retrieved January 10, 2019, from www.justiceinspectorates.gov.uk/hmicfrs/publications/state-of-policing-the-annual-assessment-of-policing-in-england-and-wales-2016.

House of Commons. (2016). *Police diversity.* First Report of Session 2016–17, HC 27. House of Commons Home Affairs Committee. London: Stationery Office.

House of Commons Committee of Public Accounts. (2017). *NHS ambulance services.* Sixty-second Report of Session 2016–17, HC 1035, April 2017.

Jacobs, G., & Keegan, A. (2018). Ethical considerations and change recipients' reactions: 'It's not all about me'. *Journal of Business Ethics, 152,* 73–90.

Jorritsma, P. Y., & Wilderom, C. (2012). Failed culture change aimed at more service provision: A Test of three agentic factors. *Journal of Organizational Change Management, 25*(3), 364–391.

Kale, P., & Singh, H. (2009, August). Managing strategic alliances: What do we know now, and where do we go from here? *Academy of Management Perspectives,* 45–62.

Knight, K. (2013). *Facing the future: Findings from the review of efficiencies and operations in fire and rescue authorities in England.* London: Department for Communities and Local Government.

Kohtamaki, M., Rabetino, R., & Moller, K. (2018). Alliance capabilities: A review and research agenda. *Industrial Marketing Management, 68,* 188–201.

Loftus, B. (2010). Police occupational culture: Classic themes, altered times. *Policing and Society: An International Journal of Research and Policy, 20*(1), 1–20.

Malin, J. R., & Hackmann, D. G. (2019). Integrative leadership and cross-sector reforms: High school career academy implementation in an urban district. *Educational Administration Quarterly, 55*(2), 189–224.

Mansfield, C. (2015). *Fire works: A collaborative way forward for the fire and rescue service.* London: New Local Government Network (NLGN).

McCalman, J., & Potter, D. (2015). *Leading cultural change: The theory and practice of successful organizational transformation.* New York: Kogan Page.

McCann, L., Wankhade, P., & Murphy, P. (2019). Conclusions. In P. Wankhade, L. McCann, & P. Murphy (Eds.), *Critical perspectives on the management and organization of emergency services.* New York: Routledge.

McCann, L., Granter, E., Hyde, P., & Hassard, J. (2013). Still blue-collar after all these years? An ethnography of the professionalization of emergency ambulance work. *Journal of Management Studies, 50*(5), 750–776.

McCann, L., Hassard, J., Granter, E., & Hyde, P. (2015). 'You can't do both: something will give': Limitations of the targets culture in managing UK health care workforces. *Human Resource Management, 54*(5), 773–791.

MIND. (2015). *Fire and rescue service: How to manage stress and anxiety.* Blue Light Programme. Retrieved January 1, 2019, from https://www.mind.org.uk/media/16771279/12195_mind-blue-light-fire-rescue-stress-booklet_new-images-2017.pdf.

National Audit Office. (2015). *Impact of funding reductions on fire and rescue services.* Retrieved February 18, 2019, from www.nao.org.uk/report/impact-of-funding-reductions-on-fire-and-rescue-services/.

National Audit Office NAO. (2011). *Transforming NHS ambulance services.* London: Stationery Office.

National Audit Office NAO. (2017). *NHS ambulance services.* HC 972, Session 2016–17. London: Stationery Office.

Ogbonna, E., & Harris, L. C. (2015). Subcultural tensions in managing organisational culture: A study of an English premier league football organisation. *Human Resource Management Journal, 25*(2), 217–232.

Page, S. (2010). Integrative leadership for collaborative governance: Civic engagement in Seattle. *The Leadership Quarterly, 21*(2), 246–263.

Paoline, E. A., & Terrill, W. (2014). *Police culture: Adapting to the strains of the job.* Durham, NC: Carolina Academic Press.

Police Federation. (2019). PFEW demand, capacity and welfare survey 2018, Headline Statistics, December 2018. Research and Policy Support Report R101/2018. Leatherhead: Police Federation.

Powell, W. W. (1998). Learning from collaboration: Knowledge and networks in the biotechnology and pharmaceutical industries. *California Management Review, 40*(3), 228–240.

Provan, K., & Kenis, P. (2008). Forms of network governance: Structure, management, and effectiveness. *Journal of Public Administration Research and Theory, 18*, 229–252.

Public Health England. (2017). *Facing the facts, shaping the future: A draft health and care workforce strategy for England to 2027.* London: Public Health England.

Silvia, C., & McGuire, M. (2010). Leading public sector networks: An empirical examination of integrative leadership behaviors. *The Leadership Quarterly, 21*(2), 264–277.

Sky News. (2018, January 11). A day in my life: Being a paramedic can feel like eat, sleep, work, repeat. *Sky News.* Retrieved February 10, 2019, from https://news.sky.com/story/what-is-it-like-being-a-paramedic-in-the-nhs-11203017.

Tehrani, N., & Hesketh, I. (2018). The role of psychological screening for emergency service responders. *International Journal of Emergency Services.* Earlycite. https://doi.org/10.1108/IJES-04-2018-002.

The King's Fund. (2015). *Making the difference: Diversity and inclusion in the NHS.* London: The King's Fund.

Thomson, A. M., & Perry, J. L. (2006). Collaboration processes: Inside the black box. *Public Administration Review, 66*, 20–32.

Vangen, S., & Huxham, C. (2012). The tangled web: Unravelling the principle of common goals in collaborations. *Journal of Public Administration Research and Theory, 22*(4), 731–760.

Vangen, S., Hayes, J. P., & Cornforth, C. (2015). Governing cross-sector interorganizational collaborations. *Public Management Review, 17*(9), 1237–1260.

Vanneste, B. S. (2016). From interpersonal to interorganisational trust: The role of indirect reciprocity. *Journal of Trust Research, 6*(1), 7–36.

Verburg, R. M., Nienaber, A.-M., Searle, R. H., Weibel, A., Den Hartog, D. N., & Rupp, D. E. (2018). The role of organizational control systems in employees' organizational trust and performance outcomes. *Group & Organization Management, 43*(2), 179–206.

Villena, H. V., Choi, T. Y., & Revilla, E. (2019). Revisiting interorganizational trust: Is more always better or could more be worse? *Journal of Management, 45*(2), 752–785.

Vize, R. (2018, March 9). NHS survey reveals staff are determined to make the best of tough conditions. *The Guardian Online.* Retrieved June 17, 2018, from https://www.theguardian.com/healthcare-network/2018/mar/09/nhs-survey-staff-determined-best-tough-conditions.

Wang, Y., & Rajagopalan, N. (2015). Alliance capabilities: Review and research agenda. *Journal of Management, 41*(1), 236–260.

Wankhade, P. (2011). Performance measurement and the UK emergency ambulance service: Unintended consequences of the ambulance response time targets. *International Journal of Public Sector Management, 24*(5), 384–402.

Wankhade, P. (2012). Different cultures of management and their relationships with organizational performance: Evidence from the UK ambulance service. *Public Money & Management, 32*(5), 381–388.

Wankhade, P. (2016). Staff perceptions and changing role of pre-hospital profession in the UK ambulance services: An exploratory study. *International Journal of Emergency Services, 5*(2), 126–144.

Wankhade, P. (2018). The crisis in NHS ambulance services in the UK: Let's deal with the 'elephants in the room'!! *Ambulance Today, 15*(1), 13–17.

Wankhade, P., & Brinkman, J. (2014). The negative consequences of culture change management: Evidence from a UK NHS ambulance service. *International Journal of Public Sector Management, 27*(1), 2–25.

Wankhade, P., Heath, G., & Radcliffe, J. (2018). Cultural change and perpetuation in organisations: Evidence from an English emergency ambulance service. *Public Management Review, 20*(6), 923–948.

Wankhade, P., McCann, L., & Murphy, P. (Eds.). (2019). *Critical perspectives on the management and organization of emergency services.* New York: Routledge.

Ward, J., & Winstanley, D. (2006). Watching the watch: The UK fire service and its impact on sexual minorities in the workplace. *Gender, Work and Organization, 13*(92), 193–219.

Williams, M. (2015, March 28). Police bullying culture deterring whistle-blowers, report warns. *The Guardian.* Retrieved June 14, 2018, from https://www.theguardian.com/uknews/2015/mar/28/police-bullying-culture-detering-whistleblowers-report-warns.

Yarnal, C. M., Dowler, L., & Hutchinson, S. (2004). Don't let the bastards see you sweat: Masculinity, public and private space, and the volunteer firehouse. *Environment and Planning A: Economy and Space, 36*(4), 685–699.

INDEX

© The Author(s) 2020

P. Wankhade, S. Patnaik, *Collaboration and Governance in the Emergency Services*,

https://doi.org/10.1007/978-3-030-21329-9